Light-Addressing and Chemical Imaging Technologies for Electrochemical Sensing

Light-Addressing and Chemical Imaging Technologies for Electrochemical Sensing

Editors

Tatsuo Yoshinobu
Michael J. Schöning

MDPI • Basel • Beijing • Wuhan • Barcelona • Belgrade • Manchester • Tokyo • Cluj • Tianjin

Editors
Tatsuo Yoshinobu
Tohoku University
Japan

Michael J. Schöning
Aachen University of Applied Sciences
Germany

Editorial Office
MDPI
St. Alban-Anlage 66
4052 Basel, Switzerland

This is a reprint of articles from the Special Issue published online in the open access journal *Sensors* (ISSN 1424-8220) (available at: https://www.mdpi.com/journal/sensors/special_issues/lacites).

For citation purposes, cite each article independently as indicated on the article page online and as indicated below:

LastName, A.A.; LastName, B.B.; LastName, C.C. Article Title. *Journal Name* **Year**, *Article Number*, Page Range.

ISBN 978-3-03943-028-4 (Hbk)
ISBN 978-3-03943-029-1 (PDF)

Contents

About the Editors

Tatsuo Yoshinobu (Ph.D.) was born in Kyoto, Japan in 1964. He received his BE, ME, and PhD degrees in electrical engineering from Kyoto University in 1987, 1989, and 1992, respectively, where he studied gas source molecular beam epitaxy of silicon carbide. In 1992, he joined the Institute of Scientific and Industrial Research, Osaka University, where he started the development of silicon-based chemical sensors. From 1999 to 2000, he was a guest scientist at the Research Centre Jülich, Germany. Since 2005, he has been a professor of electronic engineering at Tohoku University, Sendai, Japan. Since 2008, he has also been a professor at the Graduate School of Biomedical Engineering, Tohoku University.

Michael J. Schöning (Ph.D.) received his diploma degree in electrical engineering (1989) and his PhD degree in the field of semiconductor-based microsensors for the detection of ions in liquids (1993), both from Karlsruhe University of Technology. In 1989, he joined the Institute of Radiochemistry at Research Centre Karlsruhe. Since 1993, he has been with the Institute of Thin Films and Interfaces at the Research Centre Jülich (now, Institute of Biological Information Processing) and in 1999 was appointed as a full professor at Aachen University of Applied Sciences, Campus Jülich. Since 2006, he has served as the director of the Institute of Nano- and Biotechnologies (INB) at Aachen University of Applied Sciences. His main research subjects concern silicon-based chemical and biological sensors, thin-film technologies, solid-state physics, microsystem, and nano(bio-)technology.

Preface to "Light-Addressing and Chemical Imaging Technologies for Electrochemical Sensing"

Visualizing chemical components in a specimen is an essential technology in many branches of science and practical applications. While optical methods based on indicators or fluorescent dyes are widely used in biological and medical sciences, various electrochemical methods have also been developed to meet specific requirements.

A major class of electrochemical imaging techniques employs a scanning electrode to access local electrochemistry, as in the cases of scanning electrochemical microscopy (SECM) and electrochemical scanning tunneling microscopy (EC-STM). Another class of electrochemical imaging techniques relies on semiconductor devices with the capability of spatially-resolved sensing. Two types of such sensing devices have been extensively studied and applied in various fields, i.e., arrayed sensors and light-addressed sensors. An ion-sensitive field-effect transistor (ISFET) array and a charge-coupled device (CCD) ion image sensor are examples of arrayed sensors. They take advantage of semiconductor microfabrication technology to integrate a large number of sensing elements on a single chip, each representing a pixel to form a chemical image. A light-addressable potentiometric sensor (LAPS), on the other hand, has no pixel structure. A chemical image is obtained by raster-scanning the sensor plate with a light beam, which can flexibly define the position and size of a pixel. This light-addressing approach is further applied in other LAPS-inspired methods. Scanning photo-induced impedance microscopy (SPIM) realized impedance mapping and light-addressable electrodes/light-activated electrochemistry (LAE) realized local activation of Faradaic processes.

This book is a compilation of eight articles dealing with state-of-the-art technologies of light-addressing/chemical imaging devices and their application to biology and materials science. It offers graduate students, academic researchers, and industry professionals insight into different up-to-date examples. If this book gives the readers the idea of what should come next, it is our great pleasure.

Tatsuo Yoshinobu, Michael J. Schöning
Editors

Review

Light-addressable Electrodes for Dynamic and Flexible Addressing of Biological Systems and Electrochemical Reactions

Rene Welden [1,2], Michael J. Schöning [1,3], Patrick H. Wagner [2], Torsten Wagner [1,3,*]

1 Institute of Nano- and Biotechnologies (INB), Aachen University of Applied Sciences, Heinrich-Mußmann-Str. 1, 52428 Jülich, Germany
2 Laboratory for Soft Matter and Biophysics, KU Leuven, Celestijnenlaan 200D, 3001 Leuven, Belgium
3 Institute of Complex Systems (ICS-8), Research Center Jülich GmbH, 52428 Jülich, Germany
* Correspondence: torsten.wagner@fh-aachen.de; Tel.: +49-241-6009-53766

Received: 29 February 2020; Accepted: 13 March 2020; Published: 17 March 2020

Abstract: In this review article, we are going to present an overview on possible applications of light-addressable electrodes (LAE) as actuator/manipulation devices besides classical electrode structures. For LAEs, the electrode material consists of a semiconductor. Illumination with a light source with the appropiate wavelength leads to the generation of electron-hole pairs which can be utilized for further photoelectrochemical reaction. Due to recent progress in light-projection technologies, highly dynamic and flexible illumination patterns can be generated, opening new possibilities for light-addressable electrodes. A short introduction on semiconductor–electrolyte interfaces with light stimulation is given together with electrode-design approaches. Towards applications, the stimulation of cells with different electrode materials and fabrication designs is explained, followed by analyte-manipulation strategies and spatially resolved photoelectrochemical deposition of different material types.

Keywords: light-addressable electrode; light-addressable cell stimulation and photoelectrochemistry; photoelectrochemical deposition

---

1. Introduction

Charge transfer is the main task of working electrodes in electrochemistry, while they are in steady interaction with an electrolyte [1–4]. The overall application field of biosensors [5–7] extends from enzymatic biosensors [8–10] and impedimetric DNA sensors [11] to detection of action potentials of neurons with microelectrode arrays [12]. Depending on the application, electrode sizes range from macro- and micro- down to nanoelectrodes with adjustable geometries [13–15]. Regarding fabrication methods, such as thick- and thin-film technologies, several process steps are necessary. For example, in thin-film technologies, electrode design has to pass elaborated steps from photolitography to material deposition. As materials, noble metals such as platinum, gold, or silver are the most common electrode materials; however, it is also possible to use carbon [16,17], graphene [18–20], metal oxides [21–23], or conductive polymers [24–26]. Despite of their wide application field and highly developed technology standards, those electrodes are limited in their flexbility as they are usually tailored to a specific task, and due to that, they need often highly sophisticated fabrication- and read-out procedures. If a specific electrode design has to be integrated, for instance, in a lab-on-a-chip system, each single development step might require changes due to the electrode design or location. This causes a time- and resource-consuming redesign of the electrodes and the lab-on-a-chip system with an adjustment of fabrication steps. In addition, the modification of other components can have an influence on the electrode structure itself, such as the wiring of the connection.

Instead of using these classical electrode materials, semiconductors obtained attention as alternative materials for electrochemistry from first studies by Gerischer [27] and for photoinduced water splitting by Fujishima and Honda [28]. When a semiconductor is brought into contact with an electrolyte, charge carriers will exchange between the semiconductor and electrolyte until an equilibrium is reached. This is followed by a band alignment (between the valence band E_g and the conduction band E_c) in the semiconductor, where a space-charge layer is formed at the interface. When light with a suitable wavelength (photon energy > band gap energy) is absorbed by the semiconductor, electron-hole pairs can be generated. Depending on the doping, the minority charge carriers, holes (h^+) for n-type semiconductors and electrons (e^-) for p-type materials, will mainly contribute to the charge transfer at the semiconductor–electrolyte interface, where redox reactions can occur:

$$Red + h^+ \rightarrow Ox \tag{1}$$

$$Ox + e^- \rightarrow Red \tag{2}$$

In Equation (1), a reducing agent (Red) is oxidized by releasing electrons, while in Equation (2), an oxidizing agent (Ox) is reduced by gaining electrons. For example, for a n-type semiconductor, by applying an anodic potential, electrons (majority charge carriers) will move to the bulk while the holes (minority charge carriers) can perform oxidation reactions at the surface. Therefore, such a structure is called a photoanode. Opposite reactions take place at a p-type semiconductor, a photocathode. A detailed explanation of the related semiconductor physics is given in References [29,30]. The schematic working principles of photoanodes and -cathodes are shown in Figure 1.

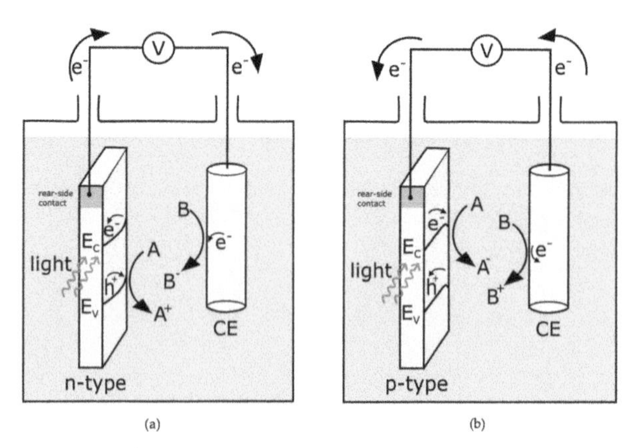

Figure 1. (**a**) n-type semiconductor as a photoanode in contact with a counter electrode (CE): At the photoanode, substance A is oxidized by the photogenerated hole, while the electron is moved to the counter electrode where it can reduce substance B. (**b**) p-type semiconductor as a photocathode in contact with a counter electrode: At the photocathode, substance A is reduced by the photogenerated electron, while the hole is moved to the counter electrode where it can oxidize substance B.

A more extensive theoretical introduction about light-addressable electrochemistry is given by Vogel et al. [31]. Well-known examples are photoelectrochemical cells [32,33] and dye-sensitized solar cells [34], to which a tremendous amount of reseach is dedicated. Nevertheless, also other fields were established based on photoelectrochemistry: photoelectrochemical sensors [35–37], for various applications, including DNA (deoxyribonucleic acid) [38], immuno [39], enzymatic [40], and heavy metal sensing [41] are known in the literature. Mainly to trigger detection, those chips are fully illuminated with a single light source and the possibility of addressing multiple analytes or areas of the sensor is disregarded. The potential of obtaining spatially resolved information on the analyte concentration was demonstrated with light-addressable potentiometric sensors [42–44]. In contrast to photoelectrochemical sensors, an insulating layer is deposited on top of the semiconductor to prevent direct charge transfer at the semiconductor–electrolyte interface. Applying a d.c. (direct current) potential, a space-charge region will be formed at the semiconductor–insulator interface. The width of this space-charge layer will change according to the ion concentration at the sensor surface. Using an intensity-modulated illumination, electron-hole pairs are continuously generated and separated in the space-charge region. Hence, the resulting alternating photocurrent is proportial to the width of the space-charge region and therefore to the ion concentration. By scanning the sensor with an appropiate optical system, spatially resolved sensor images of the analyte concentration, so-called chemical images, can be obtained. In chemical and biological systems, not only the detection of analytes or cells is of importance. An additional changing of the local environment to trigger a chemical reaction or manipulating a cell can also be significant. Especially, the localized dynamic triggering with a light source in, e.g., a lab-on-a-chip system, can be beneficial in comparison to conventional electrodes, where a sophisticated design and layout is necessary. One well-known technique which utilizes this idea is optoelectronic tweezers [45–47]. They are based on dielectrophoretic techniques by applying a.c. (alternating current) voltages between the photoconductive surface and a counter electrode, which results in nonuniform electric fields at the illuminated areas. Particles in that area might interact due to their dipoles with the electric field as dielectrophoretic forces acting on these particles. Hereby, a controlled movement of molecules or even biological cells is possible. Our review will give a closer look at three additional light-addressable electrode applications: (i) stimulation of cells, (ii) addressable photoelectrochemistry and, (iii) photoelectrochemical deposition (Figure 2).

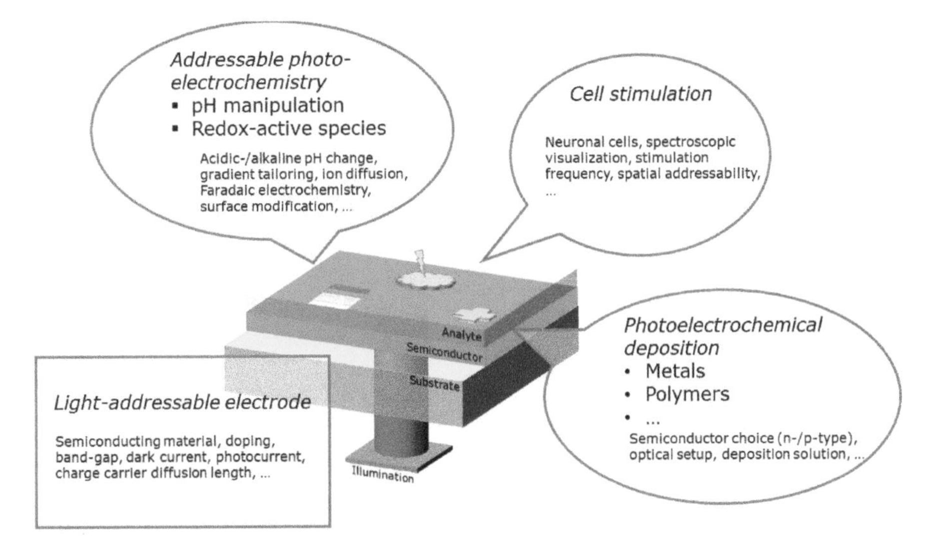

Figure 2. Overview of the presented topics in this review article: In contrast to well-known electrodes made of noble metals, light-addressable electrodes with a semiconductor electrode material can be addressed with a light source. Addressable photoelectrochemistry, cell stimulation, and photoelectrochemical deposition are introduced as possible applications.

For a successfull implementation in an experimental setup, the photoelectrode design and the optical system are the major contributors. In literature, different keywords are used for the electrode description, e.g., "photoelectrochemical", "light-induced", "light-directed", "optically directed", "photo-assisted", etc. For simplicity, the actual working electrode will be called light-addressable electrode (LAE) in this review. The design of those electrodes—see Figure 3—relies mainly on the selected semiconductor. Depending on the application, the conduction and valence band energies have to fit to the reduction and oxidation potentials of the analyte, whereby the band gap also defines the possible excitation wavelength. A detailed calculation of common band energies is given in Reference [48]. Furthermore, a sufficient dark-to-photocurrent ratio and the charge-carrier diffusion length of the semiconductor are essential to trigger a reaction only at the area of illumination. A semiconductor can be directly electrically connected (Figure 3a) or deposited as a thin film on a substrate material (Figure 3b). Often, transparent conductive oxide (TCO) glasses such as indium tin oxide (ITO) [49], fluorine-doped tin oxide (FTO) [50], or aluminium-doped zinc oxide (AZO) [51] are used as substrate materials as they allow rear-side illumination and easy electrical connection. Since LAEs are mainly used with liquid environments, the semiconductor material should be stable under the prevailing conditions. Especially the stability in alkaline or acid solutions has to be taken into account as the degradation/corrosion of the material is possible [52]. A discussion about the stability for a wide range of semiconducting materials which can be used as LAEs can be found in Reference [53,54]. Nevertheless to improve the stability, e.g., for silicon or amorphous silicon, a passivation layer can be applied (Figure 3c). The passivation layer has then to be thin enough for charge tunneling; anisotropically conductive or redox groups are required. It is also possible to integrate single noble-metal electrodes electrodes into the passivation layer to have a charge path to the analyte (Figure 3d).

In addition to the electrode design, a proper illumination source is required for flexible addressing of the LAE. In the beginning, single-focus laser spots were used for spatial illumination by mounting them onto motorized stages [55], or dimensional excitation was done by photomasks [56]. A more dynamic and flexible addressing can be achieved with Digital Micromirror Device (DMD) [57] or Ferroelectric Liquid Crystal on Silicon (FLCoS) [58] technologies. By individual switching of single pixels, both technologies (DMD and FLCoS) combine the flexibility of a scanning laser and the extensive pattern illumination of a photomask. Nevertheless, a sufficient optical focussing has to be provided to project small-scale patterns onto the electrode. In the next section, a more detailed look on the electrode design, integration of optical systems, and example applications for the stimulation of cells by the LAE will be given.

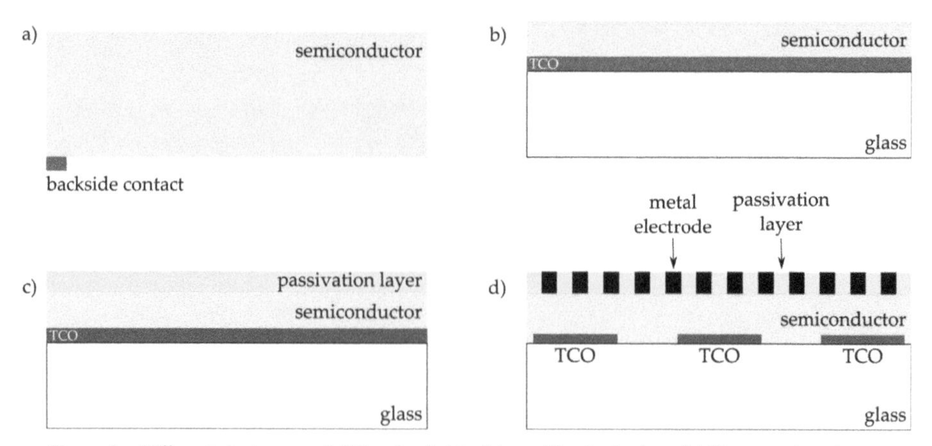

Figure 3. Different design possibilities for light-addressable electrodes: (**a**) Pure semiconductor electrically contacted, (**b**) semiconductor film deposited on a transparent conductive oxide (TCO) glass substrate for electrical connection, (**c**) deposited passivation layer for protecting the semiconductor film against environmental conditions, and (**d**) integrated metal electrodes in the passivation layer (for charge exchange) deposited on the semiconductor film.

2. Stimulation of Cells

Multielectrode arrays (MEA) are well established for manipulating cells and for recording their electrical potentials [12]. From first MEAs with 30 microelectrodes to record potentials from chicken heart cells [59] to arrays based on CMOS (complementary metal-oxide-semiconductor) technology, which have bidirectional functionality for recording and stimulation with 26,400 electrodes [60], considerable research efforts were dedicated to those electrodes. The incentive for developing new kinds of electrodes are the existing challenges. Wiring of single, fixed-positioned metal electrodes led to a limited density of electrodes with the consequence that cells lying between electrodes, or barely covering them, could only be monitored with a low signal-to-noise ratio. The high electrode density and spatial resolution, together with the good signal-to-noise ratio of nowadays CMOS technology, goes, however, along with very sophisticated fabrication and signal processing. The high costs for these MEA chips, fabricated in small numbers, make them not attractive for many potential applications. Especially, cell-culture-based applications, which need steril environments, prefer disposable systems.

Therefore, as an alternative to these technologies, light-addressable electrodes were introduced for cell stimulation by Colicos et al. [61]. In this work, photoconductive stimulation of cultured neurons was perfomed to image the green fluorescent protein (GFP) actin at individual synapses to show synaptic plasticity. A p-doped silicon wafer was used as a photoconductive material with a platinum counter electrode. For neuronal stimulation, square voltage pulses (4 V) were applied between both electrodes, while the desired photoconductive pathway was continously illuminated with a small spot (Ø = 80 μm) from an upright microscope. This experimental setup was also applied to stimulate rat hippocampal neurons in vitro. The direction of growth cones of axons was affected by triggered astrocyte Ca^{2+} waves [62]. Furthermore, different postsynaptic proteins of rat hippocampal neurons were studied at various photoconductive stimulation frequencies [63].

The technique was further improved by Campbell et al. [64]: A measurement chamber allowing for integration within an inverted confocal microscope was applied and the influence of different silicon substrates (polished p-doped silicon, porous oxidized p-doped silicon, and porous carbonized p-doped silicon) was evaluated. Rat primary cortical neurons were cultivated on the different substrates and stimulated; the Ca^{2+} response due to the stimulation was visualized with a Fluo-4 fluorescence dye. For all three substrates, the fluorescence signal of the stimulated neurons followed the stimulation frequency from the photoconductive material when pulsing the applied voltage under a constant illumination. Figure 4 shows the fluorescence signal before and after selective stimulation of neuronal cells. A single cell (marked with the crosshair) was stimulated by a laser spot, and the fluorescence increase of the adjacent cells was observed. For polished silicon, only 20% of the cells in an area of 25 μm away responded. For oxidized silicon, around 30% in a distance of 75 μm and, for carbonized silicon, 20% of the cells in a range of 200 μm away from the stimulated area are activated. This work demonstrates that, depending on the application (single-cell stimulation or stimulation of a group of neurons), different photocondutive materials can be chosen.

Silicon as an photoconductive material, which tends to oxidize without special treatment under environmental conditions, can be replaced by hydrogenated amorphous silicon (a-Si:H) to improve the spatial resolution. It is possible to deposit amorphous silicon as thin layers on a substrate material, which improves the spatial resolution due to limited lateral diffusion of the electron-hole pairs [65]. One challenge of a-Si:H is its unstable behavior in liquid environments, meaning that a passivation layer is required. The first electrodes with a-Si:H photoconductive material were introduced by Bucher et al. [66–68]. Hydrogenated amorphous silicon was deposited on a patterned (60 rows, 20 μm wide, 10 μm gap) indium tin-oxide (ITO) glass substrate, which was used for electrical connection. To protect the photoconductive film, Ti–Au or TiN electrodes were patterned on each ITO lead. As a final step, a SiO_x:C insulating layer was fabricated between the single electrodes for surface passivation. Without electrolyte, the Ti–Au electrodes showed a d.c. ohmic behavior in the range between −0.2 V to 0.2 V with a dark (without illumination) to bright (with illumination) resistance ratio of 10^7 and an a.c. impedance ratio (dark/bright) of $5 \cdot 10^6$ for the TiN electrode. In both cases, illumination of the neighbouring electrode resulted in a decrease of the dark resistance (d.c.) resp. impedance (a.c.) of the electrode under study due to light scattering. In electrolyte (physiological buffered solution), the a.c. impedance ratio (dark/bright) of the TiN electrode decreased to 30–60. Nevertheless, it was possible to record signals from cardiac myocytes cultivated on the electrodes' surface when they were illuminated.

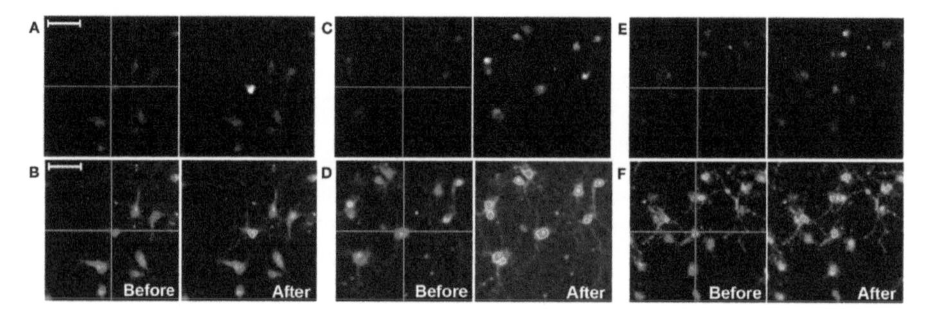

Figure 4. Fluo-4 calcium fluorescence images of rat primary cortical cells stimulated from the top-side by light-addressable electrodes with different semiconductor materials: (**A**,**B**) polished silicon, (**C**,**D**) porous oxidized silicon, and (**E**,**F**) porous carbonized silicon. On the left side of each group of four images, the grayscale image (top) and related heatmap image (bottom) before stimulation with a laser source is shown, while on the right side, the images of the stimulated neurons are depicted. The scale bar indicates 50 μm. Adapted from Reference [64] under the CC BY licence.

The deposition and patterning of single electrodes in the previous design requires a sophisticated fabrication method with sputtering, Ar-plasma-etching, plasma-enhanced chemical vapour deposition (PECVD), and CF_4-O_2-plasma-etching. The electrode design was simplified by Suzurikawa et al. by using an anisotropic passivation layer [69,70]. For electrical connection, a glass substrate with transparent SnO_2 was used with a thin a-Si:H (150 nm) photoconductor on top. To prevent dissolving, low-conductive zinc-antimonate ($ZnOSb_2O_5$)-dispersed epoxy was deposited on the a-Si:H by spin-coating and subsequent baking steps. The low water absorption of the epoxy resin prevented the culture medium from dissolving the underlying a-Si:H layer. The sheet resistance of the film was 10 MΩ/sq., and it was stable for more than two weeks in cell-culture medium. Without illumination, the charge density increased exponentially with increasing bias voltages (negative monophasic 1 ms pulses, 0–9 V), while the charge density showed a linear increase for increasing voltages under illumination. At 3 V, the best dark-to-bright charge ratio was achieved with a factor of 60. Neurons from Wistar rat embryos were plated and cultivated on the electrodes, and the stimulation was evaluated by fluorescence with Fluo-4 calcium images. For stimulation, negative voltage pulses (3 V, 1 ms) were applied between the SnO_2 layer and a counter electrode at a constant illumination (Ø = 200 μm, 800 mW/cm^2). An increase in the fluorescence signal around the illuminated area confirmed the successful stimulation of the neurons. In further studies, a light intensity between 400 and 800 mW/cm^2 was found to be sufficient for stimulation and it was possible to control stimulation pulses with a frequency of at least 500 Hz. To activate neuronal cells, a minimum charge density of 10–20 μC/cm^2 was estimated, and for single-cell stimulation, the spatial resolution was found to be 10 μm [71]. To improve the electrode performance, the dependence of the photoconductor thickness (d = 50 nm, d = 150 nm and d = 1000 nm) and two passivation layers, zinc-antimonate-dispersed epoxy (ZADE, d = 2 μm, 10 MΩ/sq.) and poly(3,4-ethylenenedioxythiophene) (PEDT, d = 0.5 μm, 4 kΩ/sq.), was analyzed. The results indicate that the performance can be further improved by a thicker photoconductive layer and by using a passivation layer which has almost the same resistivity as the photoconductive layer under illumination [72].

In another approach, Suzurikawa et al. replaced the photoconductive a-Si:H passivation sandwich structure by titanium dioxide [73]. TiO_2 anatase nanoparticles were deposited on a conductive fluorine-doped tin oxide (FTO) glass by spin-coating and subsequent sintering at different temperatures (350 °C, 500 °C and 500 °C with $TiCl_2$ treatment). For photoelectrical characterization, 1 ms negative voltage pulses were applied from 0.2 to 2 V against a Pt electrode. Without illumination, charge densities were lower than 3 $\mu C/cm^2$, which is below the suggested threshold of 10–20 $\mu C/cm^2$ for neuron stimulation. The best dark-to-bright ratio was achieved at 1.4 V with a factor of 29 (Figure 5a). Due to the mesoporous structure of the TiO_2 film and enhanced surface hydrophilicity by long-term illumination, the charge density without illumination was increased as the electrolyte solution penetrated through the film and stayed eventually in direct contact with the FTO. Furthermore, the charge density during illumination was increased with slow rising times, whereas by turning off the illumination, slow charge density decays can be observed due to the low recombination rates of photogenerated charge carriers. Nevertheless, the rising times and amplitudes of the charge density with illumination can be improved with higher sintering temperatures and $TiCl_2$ treatment. For stimulation, neurons from Wistar rat embryos were cultivated and stimulation was evaluated by Fluo-4 calcium images. Applying a 1 ms, −1.5 V voltage pulse leads to a charge density of 44 $\mu C/cm^2$ for illumination and 10 $\mu C/cm^2$ in dark regions. This was suitable for spatial resolved neuronal stimulation as seen in the increased fluorescence signals (Figure 5b).

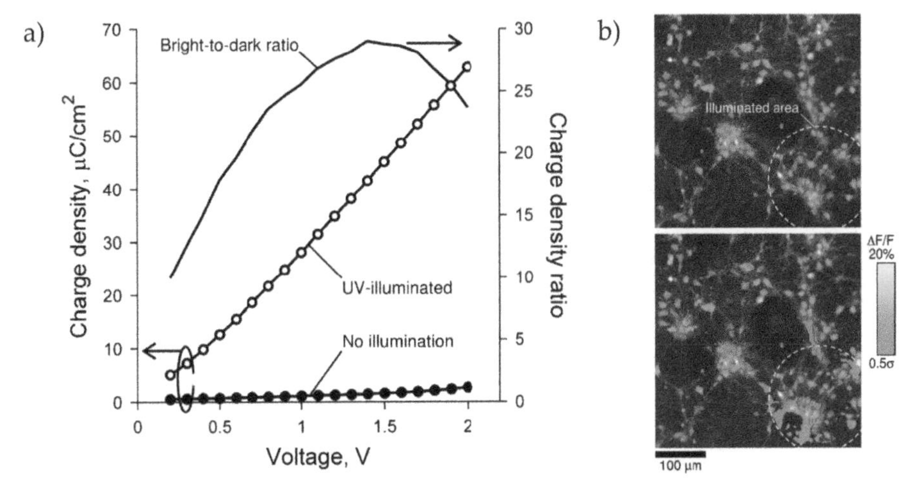

Figure 5. (**a**) Example of the charge density for a porous titanium dioxide light-addressable electrode with different, applied voltages under and without illumination. (**b**) Ca^{2+} fluorescence image of rear-side illuminated spatial stimulation of Wistar rat embryo neurons cultivated on titanium dioxide: Neurons show fluorescence changes at the illuminated areas. Reprinted from Reference [73] Copyright (2020), with permission from Elsevier.

In summary, the possibility to spatially address and trigger single cells by using light-addressable electrodes was shown in this section. For choosing the best suited material, the stability in aqueous environments, the dark-to-bright current ratio, and the spatial resolution have to be considered. If the photoconductive material is not stable in aqueous conditions, a passivation layer can be added. This in turn can influence the respective dark and photocurrents and can therefore improve or impair the dark-to-bright current ratio, which is necessary to stimulate the cell. The choice of materials can also have an influence on the spatial resolution, since the diffusion length of the generated electron-hole pairs contribute to the addressed area and therefore directly influences the addressability.

3. Addressable Photoelectrochemistry

The development of lab-on-a-chip microfluidics requires besides sensing [74,75] and flow-control elements [76] also active manipulation structures. One possible parameter to control inside those microfluidic systems is the pH value. Besides active measurements [77,78], different techniques, e.g., a.c. Faradaic reactions [79], field-enhanced water dissociation in microscale bipolar membranes [80,81], or electrolysis at electrodes inside the channel [82,83] have been developed for active manipulation of the pH value inside those microstructures. For water electrolysis, where not only gas, but also protons and hydroxide ions are generated, the anodic (Equation (3)) and cathodic (Equation (4)) reactions can be described as follows:

$$2\,H_2O \rightarrow O_2 + 4\,H^+ + 4\,e^- \tag{3}$$

$$4\,H_2O + 4\,e^- \rightarrow 2\,H_2 + 4\,OH^- \tag{4}$$

Instead of using noble metal electrodes for electrolysis, which have to be arranged in predefined locations, a more versatile method to control different pH values inside the analyzed system can be the use of photoanodes or photocathodes made of semiconductor materials. Especially, in combination with a sophisticated illumination system, a fast and spatially resolved pH changing system can be introduced. This principle was used by Hafeman et al. in his work about photoelectophoretic localization and transport (PELT) [84]. This technique enables the generation of electrical force-field traps for charged molecules by illumination on photoconductive electrodes. In combination with this, a three-dimensional positioning of amphoteric molecules by a surface pH gradient can be established. As photoanode materials, titanium dioxide ($d = 6\,\mu m$) on transparent conductive ITO glass and germanium ($d = 360\,\mu m$) photoanodes were used. Due to photoelectrolysis, the pH gradient can be controlled by illumination intensity, duration, and position. As reported, amphoteric molecules (negatively charged Bovine serum albumin) will be first attracted by the illuminated isoelectric zone. After charge exchange and more acidic conditions, they will turn the direction away from the surface. Therefore, it is possible to influence the vertical position of the molecule related to the surface in space by the pH gradient.

Suzurikawa et al. had a more detailed look at pH gradient generation on photoelectrodes [57]. In short, a-Si:H was deposited on a fluorine-doped SnO_2 (FTO) glass with a tin antimonate (ZnO/Sb_2O_5)-dispersed epoxy passivation layer. Changes of the pH value were imaged by 2′,7′-bis-(2-carboxyethyl)-5-(and-6)-carboxyfluorescein (BCECF). Voltage pulses of 400 ms from -4 V to 5 V between the photoelectrode and a counter electrode were applied. No pH changes were found for voltages between -2.5 V and 2.5 V. At -4 V, the pH change (ΔpH) was 0.65 and -0.45 at 5 V (Figure 6a). Due to the PBS (phosphate buffered saline) buffer, chlorine evolution dominates the anode reaction where protons are generated and the pH value becomes more acidic whereas hydroxide ions are directly generated at the cathode. A DMD projector was integrated in the measurement setup to generate flexible illumination patterns. Beside the pH value, also the width of the pH gradient changes with the applied voltage (Figure 6b). It was found that the pH change increases with the illumination width under same pulse durations and that the full width at half maximum (FWHM) of the ΔpH gradient profile correlates with the ΔpH peak value. By applying cathodic and anodic pulses with simultaneously defined illumination patterns, it was also possible to tune the pH gradient profiles due to neutralization reactions of hydroxide ions and protons. Furthermore, by sequentially illuminating two different areas for 400 ms, a wide-range pH gradient could be achieved when applying in the first area a cathodic potential (-3 V, $+\Delta$pH) and in the second area an anodic potential (4 V, $-\Delta$pH) (Figure 6c).

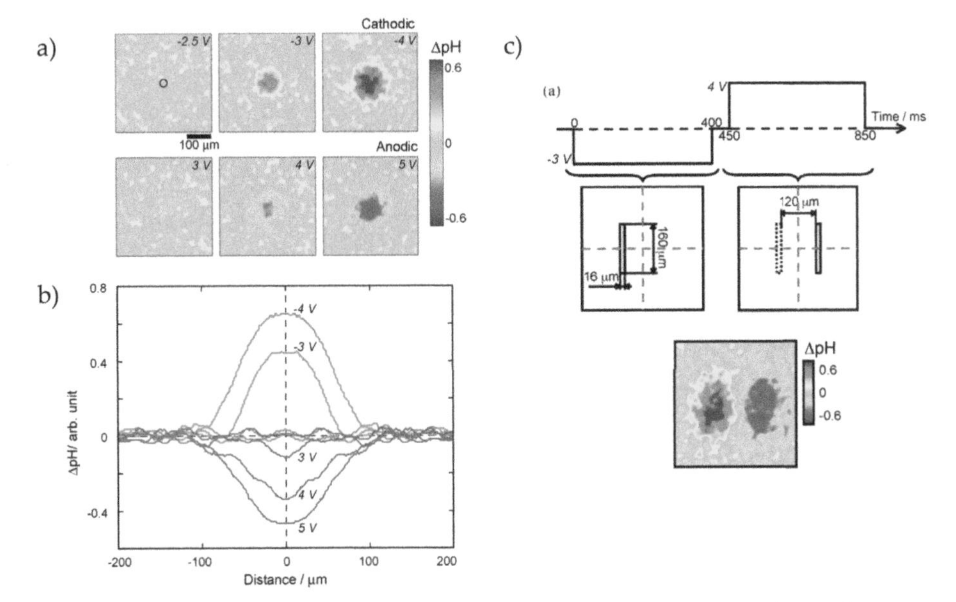

Figure 6. (**a**) Fluorescence pH images for a-Si:H with a tin antimonate (ZnO/Sb_2O_5)-dispersed epoxy passivation layer for different applied potentials with a constant rear-side illumination: For negative applied potentials, the pH changes to more alkaline value, while for positive applied potentials, the electrolyte becomes more acidic at the illuminated areas. (**b**) pH gradients for different potentials: With higher positive or negative potentials, the gradient gets wider with an absolute higher pH change. (**c**) Customized pH gradient by first applying a cathodic potential and illuminating the first area for 400 ms, where the pH gets more alkaline, followed by switching to anodic potentials for additional 400 ms together with the illumination of the seconds area. Reprinted from Reference [57] Copyright (2020), with permission from Elsevier.

This experimental setup with a DMD projector as the illumination source was also used with a glass/ITO/n-doped a-Si:H/undoped a-Si:H or glass/ITO/titanium dioxide phthalocyanine (TiOPc) structure to manipulate the pH values by photoelectrolysis for spatial deposition of calcium alginate [85,86] and chitosan [87,88]. For the a-Si:H structure, the pH change was monitored by a pH-sensitive fluoerescence dye (BCECF) diluted in DI water and an applied current density of 4 Am^{-2}. After 30 s of illumination, sharp fluoerescence images can be observed with a pH value higher than pH 6.3 while the non-illuminated areas remained at pH 5.5. After 120 s, the sharp fluoerescence image got blurred due to diffusion [87].

In other works, based on titanium dioxide [89–91], a single electrode setup was used without applied potential to photocatalytically change the pH value at illuminated regions. Water oxidation and reduction is catalyzed by photogenerated charge carriers. Due to the n-type behaviour of TiO_2, protons are generated by oxidation while electrons are transported to non-illuminated areas, where they recombine or get trapped by scavanging agents. To quantify the pH change, measurements by scanning ion-selective electrode technique (SIET) were performed to locally scan the proton generation on the electrode surface. With a nanotubular TiO_2 film, under illumination in Na_2SO_4, a pH change from pH 5.5 to pH 3.6 due to proton generation have been observed. The relaxation time of the pH value is around 40 min, while the photocurrent relaxes in 1–3 min. In the z-direction, the pH change is linearly in the range of 0–350 µm, while there is a nonlinear change from 350 to 1000 µm above the electrode surface. As depicted in Figure 7a,b, two separated electrodes (short circuited) are arranged in one chamber whereby only one electrode is illuminated. At the illuminated electrode, H^+ ions will be generated and the electrolyte gets more acidic, while at the non-illuminated electrode, the opposite reaction occurs (OH^- ion generation) and the pH is more alkaline. The arrows indicate the current density measured with the scanning vibrating electrode technique (SVET) (Figure 7b). The pH and current distribution heatmaps at the electrode areas measured by SIET and SVET are shown in Figure 7c. In the pH heatmap, the pH change at the illuminated electrode is concentrated in the area of illumination, while at the non-illuminated electrode, the pH change is distributed over the whole electrode, which is in coincidence with the current density heatmap [91]. In related work, a pH change from pH 6 to pH 5.6, pH 4.5, and pH 4 after 5 s, 1 min, and 3 min was observed [90]. This pH change can be used to modify the structural properties of pH-sensitive layer-by-layer (LbL) assembled polymer films. For the LbL structure, a positively charged poly(acrylic acid) (PAA) and a negatively charged pH-sensitive ABC triblock terpolymer, which self-assembles to core-shell-corona micelles, were deposited on titanium dioxide. Due to the pH-sensitive poly(methacrylic acid) shell, locally induced pH changes led to an increase of the thickness and a softer LbL structure.

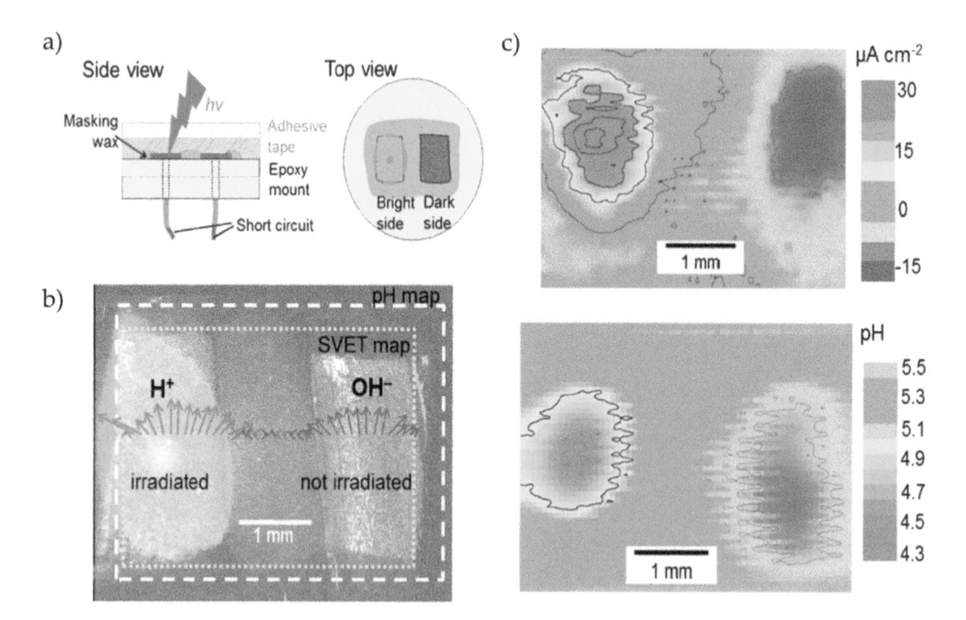

Figure 7. (**a**) Two separated titanium dioxide electrodes (n-type semiconductor) in one measurement chamber: Both electrodes are electrically shortened, while one electrode is illuminated from the top-side. (**b**) Due to the n-type behaviour, protons (H^+ ions) will be generated at the irradiated electrode, while the opposite reaction (OH^- ions) will occur at the non-illuminated electrode. The arrows indicate the current density for the anodic (red) and cathodic (blue) reactions. (**c**) Current and pH maps of both electrode regions: Positive photocurrent occurs only at the illuminated area, while at the non-irradiated electrode, a negative current is present over the complete surface. The pH changes correlate with the current map. Reprinted (adapted) with permission from Maltaneva et al. [91] Copyright (2020) American Chemical Society.

Another principle for locally induced photoelectrochemistry by light was established by Choudhury et al. who termed the technique "light-activated electrochemistry" [92]. It is based on functionalization of redox-active species on oxide-free silicon surfaces to perform light-addressed Faradaic electrochemistry on the electrode surface [93,94]. For the electrode preparation, monolayers of 1,8-nonadiyne were assembled on pretreated n- and p-doped Si(100) wafers.

The different redox-active species, ferrocene (in form of azidomethylferrocene) on n-type Si and anthraquinone (in form of 2-(azidomethyl)anthracene-9,10-dione) on p-type Si, were attached to the

1,8-nonadiyne by a copper-catalyzed azide-alkyne cycloaddition reaction (CuAAc) (Figure 8a). This work demonstrates the need to use low-doped silicon since for highly doped Si (< 0.007 Ωcm); the cyclic voltammogram shows already without illumination a behaviour similiar to a metal electrode (Figure 8b, dashed black line). In contrast to low-doped Si, for cyclic voltammograms without illumination, no redox peaks are observed for low-doped n- (8–12 Ωcm) and p-doped (1–10 Ωcm, Figure 8b, solid black line) Si. In contrast, with illumination, the ferrocene-terminated n-type Si has an $E_{1/2}$ (half-wave potential) = −140 mV *vs.* $V_{Ag/AgCl}$ and an $E_{1/2}$ = −370 mV *vs.* $V_{Ag/AgCl}$ for the p-type Si (Figure 8b, solid red line) with anthraquinone. Additionally, alternating on- and off-switching (for 200 s each) of the charge transfer at a constant applied potential (+ 0.2 V *vs.* $V_{Ag/AgCl}$) is performed without loosing activity of the redox-species after 30 cycles [92].

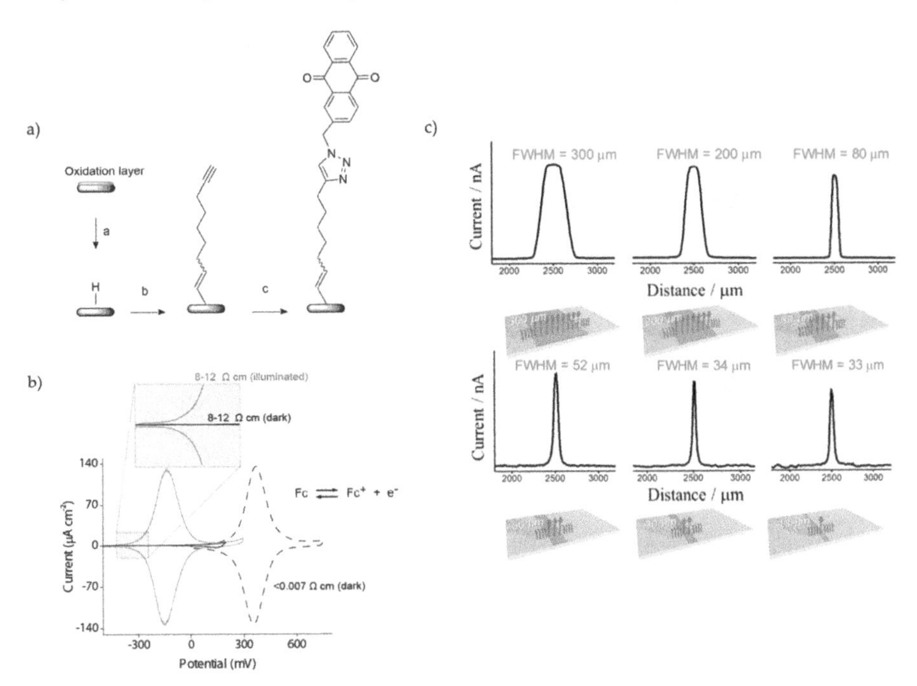

Figure 8. (**a**) Schematic of the process steps for addition of redox-active species (anthraquinone) on silicon. (**b**) Cycling voltammograms of highly doped p-type silicon(<0.007 Ωcm) without illumination compared to low-doped p-type silicon (1–10 Ωcm) without illumination (solid black line) and with rear-side illumination (solid red line). (**c**) Deposition of redox-active species with different widths (15–300 μm) to perform top-side laser line scans to determine the minimum spatial resolution by comparison with the full width at half maximum (FWHM) of the photocurrent. Part (a) is reprinted (adapted) with permission from Yang et al. [95] Copyright (2020) American Chemical Society. Part (b) is adapted from Reference [92] under the CC BY licence. Part (c) is reprinted (adapted) with permission from Yang et al. [96] Copyright (2020) American Chemical Society.

The effects of the pH value of the electrolyte and light intensity on anthraquinone deposited on low-doped p-type Si (10–20 Ωcm) were further evaluated in Reference [95]. It was shown from cyclic voltammograms that $E_{1/2}$ for modified highly p-doped Si shifts with a slope of $-58.5\,mV\,pH^{-1}$ and therefore behaves similar to a conventional metal electrode (anthraquinone monolayer on gold electrode; shift $E_{1/2} = -59.2\,mV\,pH^{-1}$), while for low-doped p-type Si , where illumination is needed for charge carrier generation, a slope of $-44.0\,mV\,pH^{-1}$ was obtained. As there was only a small change for the cathodic peak at different pH values, the anodic peak shifted by $-83.1\,mV\,pH^{-1}$. Furthermore, under a constant alkaline pH value, the anodic peak potential shows no response to an increasing light intensity, whereas the cathodic peak shifts to more positive potentials. In addition, the electron transfer rate constant for this low-doped p-type Si decreases for decreasing pH values and increases with increasing light intensities.

To evaluate the spatial resolution, anthraquinone lines of different widths (15–300 μm) were deposited on the p-doped Si. When scanning across the deposited lines, the FWHM of the related current fits to the structures down to 30 μm (Figure 8c). Furthermore, the authors stated that a top-side illumination is beneficial for the spatial resolution and additionally studied the effects of different light intensities and applied potentials [96]. Finally, as silicon as bulk material is opaque, amorphous silicon (a-Si) was deposited on ITO glass and the ferrocene was attached to 1,8-nonadiyne by the previously described CuAAc click reaction. This transparent electrode configuration is suitable for optical applications such as fluorescence imaging techniques with stimulation at the same time [97].

In conclusion, in this section, two principles for addressable photoelectrochemistry were presented. To locally change the pH value, the control of photoelectrolysis for the generation of H^+- and OH^- ions has to be taken into account. Therefore, the choice of the semiconductor material, illumination time. and applied potential has to be regulated. In another approach, redox-active species ferrocene and anthraquinone were attached to the surface of the semiconductor whereby spatially resoluted Faradaic electrochemistry can be performed.

4. Photoelectrochemical Deposition

For certain applications, the deposition of additional materials (e.g., as co-catalyst or functional layer) on semiconductor materials is desired. Besides classical electrodeposition [98,99], also light-directed deposition can be used. Using photodeposition, no bias potential is applied, and thus, the deposition depends on reductive or oxidative photo-induced processes with a sacrifical electron donor/acceptor [100]. In contrast to this, photoelectrochemical deposition is accomplished by applying an anodic or cathodic bias potential depending on the semiconductor type. Especially, the direct patterning of these materials with advanced projection technologies makes them a serious competitor to well-established technologies such as photolitographic patterning.

For metal deposition, metal ions from the solution are reduced by photogenerated electrons in the conduction band upon illumination of the semiconductor, which is described by Equation (5):

$$M^{n+} + ne^-_{surf} \rightarrow M_{surf} \quad (5)$$

where M^{n+} is a metal ion of charge n in solution, e^- is a photogenerated electron, and M_{surf} is metal atoms on the surface.

For example, gold, copper, and nickel were deposited on p-Si and p-GaAs using standard plating solutions [101]. Photoplating currents range from 20 to 250 μA for cathodically applied potentials (-1 to -2 V *vs.* V_{SCE}) (SCE: saturated calomel electrode). As long as no overpotential is applied, deposition only occurs during illumination. Gold was also used on three-dimensional silicon microwires by applying -1.25 V *vs.* $V_{Ag/AgCl}$ in 0.01 M $HAuCl_4$ solution [102]. Anisotropic Au deposition was demonstrated by using cylindrical microwires, whereby the gold position on the microwires mainly depends on the illumination wavelength. Furthermore, using noncylindrical microwires in combination with computer simulations (Finite Difference Time Domain Method), there was a correlation between the plating position and the concentration of generated charge carriers in the semiconductor.

For the reduction of metal ions to metals, normally p-type semiconductors are used, while for metal oxides, a n-type semiconductor is necessary [103]. Nevertheless, Au, Cu, Pt, and Ag could be deposited on n-type titanium dioxide in the presence of alcohols [104]. A TiO_2 photoanode was biased at 0.2 V *vs.* V_{SCE} in 0.1 M Na_2SO_4, 2 mM metal salt (H_2PtCl_6, $HAuCl_4$, $AgNO_3$, or $CuCl_2$) and 10 vol% of different alcohols. As explained in Reference [104], the reaction is based on the "current doubling effect", where holes oxidize alcohols to radicals, which inject another electron in the conduction band (mostly trap states). Furthermore, they explained that those surface electron traps can then reduce the metal ions.

The biggest benefit of photoelectrochemical deposition is the patterning of the desired material. One factor affecting the resolution of the structured material is the charge-carrier diffusion length within the semiconductor. Therefore, Pt, Au, and Ni–Mo were deposited on amorphous silicon, which was passivated with a thin (d = 1–2 nm) SiO_x passivation layer. As an optical system, a DMD projector with a 2× objective lens was used. For Pt (1 mM Na_2PtCl_6 in 0.5 M Na_2SO_4), a cathodic potential (−0.4 V vs. $V_{Ag/AgCl}$) was applied for 180 s. The Pt image has a minimal resolution of 100 µm. Additionally, Ni–Mo (130 mM $Ni(SO_3NH_2)_2$, 50 mM H_3BO_3, and 2 mM $NaMoO_4$ with pH adjusted to pH 4.0) was patterned at potentials of −0.9 V for 10 s followed by a 5 s open circuit potential (+ 0.35 V) to replenish the precursor ions [105]. Similiar to amorphous silicon, also hematite (α-Fe_2O_3) fullfills the requirements for a high spatial resolution through a hole diffusion length of 1–2 nm [106]. Co–Pi was deposited on hematite in different studies [106–108]. The oxidation of Co^{2+} to Co^{3+}, triggered by a photogenerated hole was done in a 0.5 mM $Co(NO_3)_2$, 0.1 M potassium phosphate buffer at pH 7. Applying alternating voltage pulses of 0.4 V and 0 V (vs. $V_{Ag/AgCl}$) for 5 s in combination with a DMD projector leads to Co–Pi structures with radii from 20 to 200 µm, whereas for all structures, the deviation of the accomplished pattern to the illuminated area is less then 1 µm. As a last example, cuprous oxide (Cu_2O, 50 mM $CuSO_4$, 0.5 M K_2SO_4, and a specific amount of KCl) was deposited on amorphous silicon pasivated by a 1,8-nonadiyne layer. Pattern projection was performed by a FLCoS micromirror device. It was shown that, by changing the applied bias only, particle patterns with different cubicities can be achieved [58]. In further work, effects to change the deposited particle shape, e.g., by variation of pixel density, chloride concentration and potential were studied. As a result, by first illuminating one part of the final pattern with a defined pixel gradient followed by illuminating the second part with a different Cl^- concentration, areas with different Cu_2O shapes can be implemented (Figure 9a) [109].

Besides metals, also polymers such as polystyrene [110], polyaniline [111], or polypyrrole [55,56,112–115] were sucessfully tested. Since polymerizations usually include oxidation reactions, mostly n-type semiconductors are used. In the first studies by Okano et al., polypyrrole was deposited on titanium dioxide in a three-electrode cell using a 0.1 M monomer solution of pyrrole in 0.1 M Na_2SO_4 with an applied potential 0.5 V *vs.* V_{SCE}. Masking the electrode leads to a minimum achievable line width of 45 µm [114]. It should be noted that the achievable resolution is limited by a nonsymmetrical polymerization of pyrrole [104] and that an enhanced charge transfer leads to blurring of the pattern [114]. As patterning was mostly done with shadow masks or movable lasers, it can also be spatially illuminated, e.g., by a DMD projector on a TiO_2 LAE [116]. In a recently performed experiment by our group, it was shown that, after ten potentiodynamic cycles from − 0.2 to 1.2 V *vs.* V_{Pt} under illumination in 0.1 M pyrrole (0.1 M Na_2SO_4), the letters "TiO2" were successfully patterned on the surface (Figure 9b). Subsequently, the influence of the applied potential related to the contribution of electrochemical and photoelectrochemical polymerization [111,117] and the control of polymer homogeneity and thickness by the wavelength [115] were further analyzed. Additionally, the polymerization was performed on silicon [55], ZnO [113], and WO_3 [111].

The light-induced pH change introduced in Section 3 can also be used for the deposition of pH-sensitive materials. For calcium alginate gels, the decreased pH leads to a local release of calcium ions (Ca^{2+}) from $CaCO_3$ with a subsequent cross-linking with alginate (from sodium alginate) to form a calcium alginate gel. This gel can, for example, be used to entrap cells or enzymes [118]. On undoped amorphous silicon, calcium alginate structures with diameters down to 100 µm can be accomplished as depicted in Figure 10a [85]. It was found that the ratio between the illumination patterns and actually achieved diameter (deposition ratio) varies with illumination time, alginate and $CaCO_3$ concentration, and the applied voltage. Furthermore, fibroblasts were added to the solution and, under certain conditions, the cells had a viability of more than 90%. Deposition of calcium alginate was also performed on TiOPc [86]. Different geometries were patterned, ranging from connected calcium alginate blocks and stacked layers to vessel-shaped structures by sequential illumination of the electrode.

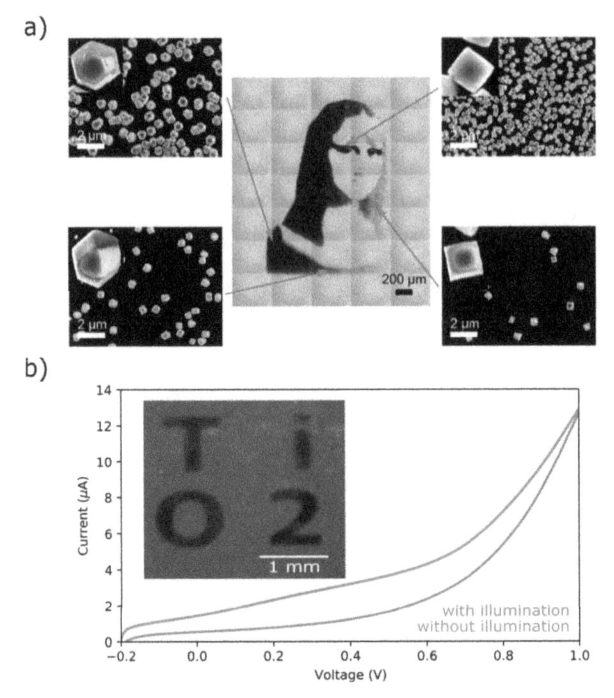

Figure 9. (**a**) Partial deposition of Cu_2O by top-side illumination with a Ferroelectric Liquid Crystal on Silicon (FLCoS) display under various conditions to influence the particles' shape and cubicity. (**b**) Snippet of the first potentiodynamic cycle with and without rear-side illumination during the polymerization of pyrrole on TiO_2. In the inset, the deposited "TiO2" letters on the surface is shown. Part (a) is used with permission from Vogel et al. [109] Copyright (2020) Wiley.

In contrast to calcium alginate, the polysaccharide chitosan can form a hydrogel for a solution pH above pH 6.3 at illuminated areas on an amorphous silicon photocathode [87,88]. Different shapes and geometries with a minimum resolution of 40 µm were implemented. Similar to alginate, increasing the deposition time and the current density leads to an increase of the deposition ratio due to the OH^- ion diffusion. As shown in Figure 10b, multiple patterns were consecutively deposited with one part of the hydrogels containing glucose oxidase, peroxidase, and amplex red as a fluorescence indicator (circular shape) and the other part containing alcohol oxidase, peroxidase, and amplex red (quadratic shape). By adding glucose or ethanol, the fluorescence intensity of the particular structure increases proportianally to the concentration. Due to the different shapes, an optical readout of glucose and ethanol sensing was possible.

Upon pH changes, the light-addressed binding control of (poly)-histidine-tagged (His-Tag) proteins was also introduced [89]. A layer-by-layer composition of polystyrene sulfonate (PSS) and nickel-nitrilotriacetic (NTA) for the His-Tag binding on titanium dioxide was described. Since only the His-Tag is pH-sensitive, it can be locally adsorbed and released by pH changes triggered by illumination.

For simultaneous DNA analysis, the defined positioning of DNA microarrays on the substrate is crucial. Therefore, the spatial arrangement of DNA oligonucleotides by photoelectrophoretic transport was done with a layer structure consisting of n-type silicon (amorphous silicon), Mn_2O_3 passivation, and agarose-streptavidin functional layer in L-histidine solution [119]. Upon illumination, biotinylated (for streptavidin binding) and fluorescence-marked DNA oligonucleotides were bound in the illuminated areas. As for n-type silicon, the results were not reproducible; a better spatial deposition was achieved with amorphous silicon. Additionally, successfull hybridization with target DNA strands was also shown. In another approach, the electrode structure was modified and amorphous silicon was deposited on ITO glass with structured platinum pads on top. To increase the surface area, an additional porous glass layer was deposited [120]. Spatially resolved detritlyation of surface-bounded 4,4-dimethoxytrityl protecting groups by photoelectrochemically generated protons led to a subsequent hybridization of oligonucleotides.

With the previously described semiconductor electrodes, it is possible to pattern a considerable amount of materials summarized in Table 1. Depending on whether a reduction or oxidation is neccesary for deposition, a n- or p-type semiconductor has to be chosen. Furthermore, to achieve a defined pattern, the diffusion length of the semiconductor and the blurring of the deposited material with time have to be taken into account.

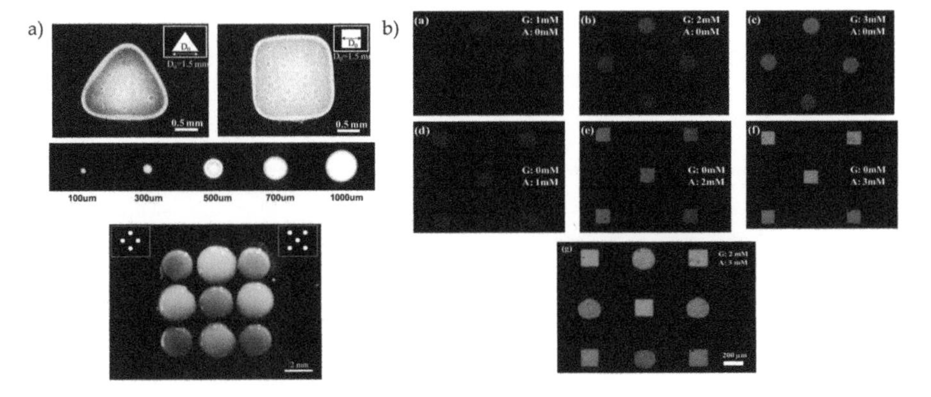

Figure 10. (**a**) Deposition of calcium alginate gels by rear-side illumination of hydrogenated amorphous silicon: In the top image, calcium alginate gels are deposited with different geometries and sizes. In the bottom image, the gels are deposited sequentially with two different micropatterns. (**b**) Patterning of chitosan with different geometries by rear-side illumination of an amorphous silicon photocathode: In the circular structure, glucose oxidase, peroxidase, and amplex red is added, while in the quadraric structure, alcohol oxidase is added instead of glucose oxidase. Adding ethanol or glucose increases the fluorescence intensity of the particular structure. Part (a) is used with permission from Huang et al. [85] Copyright (2020) AIP. Part (b) is adapted from Reference [87] under the CC BY licence.

Table 1. Summary of feasible materials which can be photoelectrochemically deposited on semiconductor electrodes.

Type of Material	Deposited Material	Type of Semiconductor	References
Metal (oxide)	Au	p-Si	[101,102]
	Au	p-GaAs	[101,121]
	Ag	p-GaP	[103]
	Cu, Ni	p-Si	[101]
	Cu, Ni	p-GaAs	[101]
	Au, Cu, Pt, Ag	TiO_2	[104]
	Pt, Au, Ni–Mo	a:Si	[105]
	Co–Pi	α-Fe_2O_3	[106–108]
	Cu_2O	a:Si	[105]

Table 1. *Cont.*

Type of Material	Deposited Material	Type of Semiconductor	References
Polymer	Polystyrene	TiO_2	[110]
	Polyaniline	WO_3	[111]
	Polyaniline	TiO_2	[111]
	Polypyrrol	TiO_2	[56,111,112,114,115,122]
	Polypyrrol	n-Si	[55,92]
	Polypyrrol	ZnO	[113]
Others	Calcium alginate gels	a:Si	[110]
	Calcium alginate gels	TiOPc	[85,86]
	Chitosan	a:Si	[87,88]
	Polyhistidine-tagged proteins	TiO_2	[89]
	DNA oligonucleotids	n-Si	[119]
	DNA oligonucleotids	a:Si	[120]

5. Conclusions

This review gives an overview on various applications for light-addressable electrodes. The basic principles of semiconductor electrochemistry were elucidated, and design strategies were explained (Section 1). For biological applications, the stimulation of neuronal cells cultivated on different semiconductor substrates was presented (Section 2). To address single cells, the material choice and design of the electrode have an influence on the spatial addressability since, depending on these factors, additional cells were stimulated outside the excited area. Area-selective photoelectrochemistry by spatially changing the analyte pH value and designing pH gradients and the possibility of light-activated electrochemistry have been depicted in Section 3. As a last topic, the possibility of structuring metals, polymers, and macromolcules with modern illumination strategies is shown. Due to the fact that the electrodes had mostly the same structure, the various applications in Section 4 can be addressed with the same setup. This opens the feasibility to combine different fields or to embed similar technologies such as the optoelectronic tweezers with the stimulation of cells. To conclude this review, with light-addressable electrodes, complex electrode structures, and a dynamic triggering of reactions such as the previously introduced cell stimulation, pH manipulation or material deposition can be achieved with a single electrode structure combined with an illumination system. This has the benefit to be easily adaptable to new applications or changes in later experimental stages in contrast to commonly used electrodes, where modifications are often followed by a redesign of the electrode.

Author Contributions: The conceptualization for this review was done by M.J.S., T.W., and R.W. The investigation and writing was conducted by R.W. Supervision, dicussion of the review, and editing of the presented work were performed by M.J.S., P.H.W., and T.W. All authors have read and agreed to the published version of the manuscript.

Funding: This research project was partially funded by the German Federal Ministry of Education and Research (BMBF) within the research frame of "NanoMatFutur": 13N12585.

Conflicts of Interest: The authors declare no conflict of interest. The funding sponsors had no role in the design of the study; in the collection, analysis, or interpretation of date; in the writing of the manuscript; and in the decision to publish the results.

References

1. Bard, A.J.; Faulkner, L.R.; Leddy, J.; Zoski, C.G. *Electrochemical Methods: Fundamentals and Applications*, 2nd ed.; Wiley: New York, NY, USA, 1980.

2. Morrison, S.R. *Electrochemistry at Semiconductor and Oxidized Metal Electrodes*; Plenum Press: New York, NY, USA, 1980.
3. Bruce, P.G. *Solid State Electrochemistry*, 5th ed.; Cambridge University Press: Cambridge, UK, 1997.
4. Schmickler, W.; Santos, E. *Interfacial Electrochemistry*; Springer Science & Business Media: New York, NY, USA, 2010.
5. Turner, A.; Karube, I.; Wilson, G. *Biosensors: Fundamentals and Applications*; Oxford University Press: Oxford, UK, 1987.
6. Bakker, E.; Telting-Diaz, M. Electrochemical sensors. *Anal. Chem.* **2002**, *74*, 2781–2800. [CrossRef] [PubMed]
7. Ronkainen, N.J.; Halsall, H.B.; Heineman, W.R. Electrochemical biosensors. *Chem. Soc. Rev.* **2010**, *39*, 1747–1763. [CrossRef]
8. Pilas, J.; Selmer, T.; Keusgen, M.; Schöning, M.J. Screen-printed carbon electrodes modified with graphene oxide for the design of a reagent-free NAD^+-dependent biosensor array. *Anal. Chem.* **2019**, *91*, 15293–15299. [CrossRef]
9. Molinnus, D.; Poghossian, A.; Keusgen, M.; Katz, E.; Schöning, M.J. Coupling of biomolecular logic gates with electronic transducers: From single enzyme logic gates to sense/act/treat chips. *Electroanalysis* **2017**, *29*, 1840–1849. [CrossRef]
10. Molinnus, D.; Hardt, G.; Käver, L.; Willenberg, H.S.; Kröger, J.C.; Poghossian, A.; Keusgen, M.; Schöning, M.J. Chip-based biosensor for the detection of low adrenaline concentrations to support adrenal venous sampling. *Sens. Actuators B* **2018**, *272*, 21–27. [CrossRef]
11. Riedel, M.; Kartchemnik, J.; Schöning, M.J.; Lisdat, F. Impedimetric DNA detection–steps forward to sensorial application. *Anal. Chem.* **2014**, *86*, 7867–7874. [CrossRef] [PubMed]
12. Spira, M.E.; Hai, A. Multi-electrode array technologies for neuroscience and cardiology. *Nat. Nanotechnol.* **2013**, *8*, 83–94. [CrossRef]
13. Forster, R.J. Microelectrodes: New dimensions in electrochemistry. *Chem. Soc. Rev.* **1994**, *23*, 289–297. [CrossRef]
14. Arrigan, D.W. Nanoelectrodes, nanoelectrode arrays and their applications. *Analyst* **2004**, *129*, 1157–1165. [CrossRef]
15. Rothe, J.; Frey, O.; Stettler, A.; Chen, Y.; Hierlemann, A. Fully integrated CMOS microsystem for electrochemical measurements on 32× 32 working electrodes at 90 frames per second. *Anal. Chem.* **2014**, *86*, 6425–6432. [CrossRef]
16. Wang, J. Carbon-nanotube based electrochemical biosensors: A review. *Electroanalysis* **2005**, *17*, 7–14. [CrossRef]
17. McCreery, R.L. Advanced carbon electrode materials for molecular electrochemistry. *Chem. Rev.* **2008**, *108*, 2646–2687. [CrossRef] [PubMed]
18. Shao, Y.; Wang, J.; Wu, H.; Liu, J.; Aksay, I.A.; Lin, Y. Graphene based electrochemical sensors and biosensors: A review. *Electroanalysis* **2010**, *22*, 1027–1036. [CrossRef]
19. Chen, D.; Tang, L.; Li, J. Graphene-based materials in electrochemistry. *Chem. Soc. Rev.* **2010**, *39*, 3157–3180. [CrossRef]
20. Huang, X.; Zeng, Z.; Fan, Z.; Liu, J.; Zhang, H. Graphene-based electrodes. *Adv. Mater.* **2012**, *24*, 5979–6004. [CrossRef]
21. Solanki, P.R.; Kaushik, A.; Agrawal, V.V.; Malhotra, B.D. Nanostructured metal oxide-based biosensors. *NPG Asia Mater.* **2011**, *3*, 17–24. [CrossRef]
22. Hahn, Y.B.; Ahmad, R.; Tripathy, N. Chemical and biological sensors based on metal oxide nanostructures. *Chem. Commun.* **2012**, *48*, 10369–10385. [CrossRef]
23. Jiang, J.; Li, Y.; Liu, J.; Huang, X.; Yuan, C.; Lou, X.W. Recent advances in metal oxide-based electrode architecture design for electrochemical energy storage. *Adv. Mater.* **2012**, *24*, 5166–5180. [CrossRef] [PubMed]
24. Cho, S.I.; Lee, S.B. Fast electrochemistry of conductive polymer nanotubes: synthesis, mechanism, and application. *Acc. Chem. Res.* **2008**, *41*, 699–707. [CrossRef]
25. Nambiar, S.; Yeow, J.T. Conductive polymer-based sensors for biomedical applications. *Biosens. Bioelectron.* **2011**, *26*, 1825–1832. [CrossRef]
26. Naveen, M.H.; Gurudatt, N.G.; Shim, Y.B. Applications of conducting polymer composites to electrochemical sensors: A review. *Appl. Mater. Today* **2017**, *9*, 419–433. [CrossRef]

27. Gerischer, H. Electrochemical behavior of semiconductors under illumination. *J. Electrochem. Soc.* **1966**, *113*, 1174–1182. [CrossRef]
28. Fujishima, A.; Honda, K. Electrochemical photolysis of water at a semiconductor electrode. *Nature* **1972**, *238*, 37–38. [CrossRef] [PubMed]
29. Sze, S.M.; Ng, K.K. *Physics of Semiconductor Devices*; John Wiley & Sons: Hoboken, NJ, USA, 2006.
30. Neamen, D.A. *Semiconductor Physics and Devices: Basic Principles*; McGraw-Hill: New York, NY, USA, 2012.
31. Vogel, Y.B.; Gooding, J.J.; Ciampi, S. Light-addressable electrochemistry at semiconductor electrodes: Redox imaging, mask-free lithography and spatially resolved chemical and biological sensing. *Chem. Soc. Rev.* **2019**, *48*, 3723–3739. [CrossRef] [PubMed]
32. Grätzel, M. Photoelectrochemical cells. *Nature* **2001**, *414*, 338–344. [CrossRef]
33. Van de Krol, R. Principles of photoelectrochemical cells. In *Photoelectrochemical Hydrogen Production*; Springer: Boston, MA, USA, 2012; pp. 13–67.
34. Grätzel, M. Dye-sensitized solar cells. *J. Photochem. Photobiol. C* **2003**, *4*, 145–153. [CrossRef]
35. Devadoss, A.; Sudhagar, P.; Terashima, C.; Nakata, K.; Fujishima, A. Photoelectrochemical biosensors: New insights into promising photoelectrodes and signal amplification strategies. *J. Photochem. Photobiol. C* **2015**, *24*, 43–63. [CrossRef]
36. Zhao, W.W.; Xu, J.J.; Chen, H.Y. Photoelectrochemical bioanalysis: The state of the art. *Chem. Soc. Rev.* **2015**, *44*, 729–741. [CrossRef] [PubMed]
37. Pang, H.; Zang, Y.; Fan, J.; Yun, J.; Xue, H.G. Current advances in semiconductor nanomaterials-based photoelectrochemical biosensing. *Chem. Eur. J.* **2018**, *24*, 14010–14027.
38. Zhao, W.W.; Xu, J.J.; Chen, H.Y. Photoelectrochemical DNA biosensors. *Chem. Rev.* **2014**, *114*, 7421–7441. [CrossRef]
39. Zhao, W.W.; Xu, J.J.; Chen, H.Y. Photoelectrochemical immunoassays. *Anal. Chem.* **2017**, *90*, 615–627. [CrossRef] [PubMed]
40. Zhao, W.W.; Xu, J.J.; Chen, H.Y. Photoelectrochemical enzymatic biosensors. *Biosens. Bioelectron.* **2017**, *92*, 294–304. [CrossRef] [PubMed]
41. Zhao, W.W.; Xu, J.J.; Chen, H.Y. Photoelectrochemical detection of metal ions. *Analyst* **2016**, *141*, 4262–4271. [CrossRef] [PubMed]
42. Wagner, T.; Werner, C.F.; Miyamoto, K.i.; Schöning, M.J.; Yoshinobu, T. Development and characterisation of a compact light-addressable potentiometric sensor (LAPS) based on the digital light processing (DLP) technology for flexible chemical imaging. *Sens. Actuators B* **2012**, *170*, 34–39. [CrossRef]
43. Yoshinobu, T.; Miyamoto, K.i.; Wagner, T.; Schöning, M.J. Recent developments of chemical imaging sensor systems based on the principle of the light-addressable potentiometric sensor. *Sens. Actuators B* **2015**, *207*, 926–932. [CrossRef]
44. Yoshinobu, T.; Miyamoto, K.i.; Werner, C.F.; Poghossian, A.; Wagner, T.; Schöning, M.J. Light-addressable potentiometric sensors for quantitative spatial imaging of chemical species. *Annu. Rev. Anal. Chem.* **2017**, *10*, 225–246. [CrossRef]
45. Chiou, P.Y.; Ohta, A.T.; Wu, M.C. Massively parallel manipulation of single cells and microparticles using optical images. *Nature* **2005**, *436*, 370–372. [CrossRef]
46. Ohta, A.T.; Chiou, P.Y.; Han, T.H.; Liao, J.C.; Bhardwaj, U.; McCabe, E.R.; Yu, F.; Sun, R.; Wu, M.C. Dynamic cell and microparticle control via optoelectronic tweezers. *J. Microelectromech. Syst.* **2007**, *16*, 491–499. [CrossRef]
47. Jamshidi, A.; Pauzauskie, P.J.; Schuck, P.J.; Ohta, A.T.; Chiou, P.Y.; Chou, J.; Yang, P.; Wu, M.C. Dynamic manipulation and separation of individual semiconducting and metallic nanowires. *Nat. Photonics* **2008**, *2*, 85–89. [CrossRef]
48. Chen, S.; Wang, L.W. Thermodynamic oxidation and reduction potentials of photocatalytic semiconductors in aqueous solution. *Chem. Mater.* **2012**, *24*, 3659–3666. [CrossRef]
49. Mryasov, O.; Freeman, A.J. Electronic band structure of indium tin oxide and criteria for transparent conducting behavior. *Phys. Rev. B* **2001**, *64*, 233111. [CrossRef]
50. Rakhshani, A.; Makdisi, Y.; Ramazaniyan, H. Electronic and optical properties of fluorine-doped tin oxide films. *J. Appl. Phys.* **1998**, *83*, 1049–1057. [CrossRef]

51. Van Gompel, M.; Conings, B.; Jiménez Monroy, K.; D 'Haen, J.; Gilissen, K.; D 'Olieslaeger, M.; Van Bael, M.; Wagner, P. Preparation of epitaxial films of the transparent conductive oxide Al: ZnO by reactive high-pressure sputtering in Ar/O_2 mixtures. *Phys. Status Solidi (a)* **2013**, *210*, 1013–1018. [CrossRef]
52. Bae, D.; Seger, B.; Vesborg, P.C.; Hansen, O.; Chorkendorff, I. Strategies for stable water splitting via protected photoelectrodes. *Chem. Soc. Rev.* **2017**, *46*, 1933–1954. [CrossRef] [PubMed]
53. Sivula, K.; Van De Krol, R. Semiconducting materials for photoelectrochemical energy conversion. *Nat. Rev. Mater.* **2016**, *1*, 15010. [CrossRef]
54. Jiang, C.; Moniz, S.J.; Wang, A.; Zhang, T.; Tang, J. Photoelectrochemical devices for solar water splitting—Materials and challenges. *Chem. Soc. Rev.* **2017**, *46*, 4645–4660. [CrossRef] [PubMed]
55. Yoneyama, H.; Kitayama, M. Photocatalytic deposition of light-localized polypyrrole film pattern on n-type silicon wafers. *Chem. Lett.* **1986**, *15*, 657–660. [CrossRef]
56. Okano, M.; Itoh, K.; Fujishima, A.; Honda, K. Generation of organic conducting patterns on semiconductors by photoelectrochemical polymerization of pyrrole. *Chem. Lett.* **1986**, *15*, 469–472. [CrossRef]
57. Suzurikawa, J.; Nakao, M.; Kanzaki, R.; Takahashi, H. Microscale pH gradient generation by electrolysis on a light-addressable planar electrode. *Sens. Actuators B* **2010**, *149*, 205–211. [CrossRef]
58. Vogel, Y.B.; Gonçales, V.R.; Gooding, J.J.; Ciampi, S. Electrochemical microscopy based on spatial light modulators: A projection system to spatially address electrochemical reactions at semiconductors. *J. Electrochem. Soc.* **2018**, *165*, H3085–H3092. [CrossRef]
59. Thomas, C., Jr.; Springer, P.; Loeb, G.; Berwald-Netter, Y.; Okun, L. A miniature microelectrode array to monitor the bioelectric activity of cultured cells. *Exp. Cell Res.* **1972**, *74*, 61–66. [CrossRef]
60. Ballini, M.; Müller, J.; Livi, P.; Chen, Y.; Frey, U.; Stettler, A.; Shadmani, A.; Viswam, V.; Jones, I.L.; Jäckel, D.; et al. A 1024-channel CMOS microelectrode array with 26,400 electrodes for recording and stimulation of electrogenic cells in vitro. *IEEE J. Solid-State Circuits* **2014**, *49*, 2705–2719. [CrossRef] [PubMed]
61. Colicos, M.A.; Collins, B.E.; Sailor, M.J.; Goda, Y. Remodeling of synaptic actin induced by photoconductive stimulation. *Cell* **2001**, *107*, 605–616. [CrossRef]
62. Hung, J.; Colicos, M.A. Astrocytic Ca^2+ waves guide CNS growth cones to remote regions of neuronal activity. *PLoS ONE* **2008**, *3*, e3692. [CrossRef]
63. Gutiérrez, R.C.; Flynn, R.; Hung, J.; Kertesz, A.C.; Sullivan, A.; Zamponi, G.W.; El-Husseini, A.; Colicos, M.A. Activity-driven mobilization of post-synaptic proteins. *Eur. J. Neurosci.* **2009**, *30*, 2042–2052. [CrossRef] [PubMed]
64. Campbell, J.; Singh, D.; Hollett, G.; Dravid, S.M.; Sailor, M.J.; Arikkath, J. Spatially selective photoconductive stimulation of live neurons. *Front. Cell. Neurosci.* **2014**, *8*, 1–9. [CrossRef]
65. Moritz, W.; Yoshinobu, T.; Finger, F.; Krause, S.; Martin-Fernandez, M.; Schöning, M.J. High resolution LAPS using amorphous silicon as the semiconductor material. *Sens. Actuators B* **2004**, *103*, 436–441. [CrossRef]
66. Bucher, V.; Brunner, B.; Leibrock, C.; Schubert, M.; Nisch, W. Electrical properties of a light-addressable microelectrode chip with high electrode density for extracellular stimulation and recording of excitable cells. *Biosens. Bioelectron.* **2001**, *16*, 205–210. [CrossRef]
67. Bucher, V.; Schubert, M.; Kern, D.; Nisch, W. Light-addressed sub-μm electrodes for extracellular recording and stimulation of excitable cells. *Microelectron. Eng.* **2001**, *57*, 705–712. [CrossRef]
68. Bucher, V.; Brugger, J.; Kern, D.; Kim, G.M.; Schubert, M.; Nisch, W. Electrical properties of light-addressed sub-μm electrodes fabricated by use of nanostencil-technology. *Microelectron. Eng.* **2002**, *61*, 971–980. [CrossRef]
69. Suzurikawa, J.; Takahashi, H.; Takayama, Y.; Warisawa, S.; Mitsuishi, M.; Nakao, M.; Jimbo, Y. Light-addressable planar electrode with hydrogenated amorphous silicon and low-conductive passivation layer for stimulation of cultured neurons. In Proceedings of the 2006 International Conference of the IEEE Engineering in Medicine and Biology Society, New York, NY, USA, 30 August–3 September 2006; pp. 648–651.
70. Suzurikawa, J.; Takahashi, H.; Kanzaki, R.; Nakao, M.; Takayama, Y.; Jimbo, Y. Light-addressable electrode with hydrogenated amorphous silicon and low-conductive passivation layer for stimulation of cultured neurons. *Appl. Phys. Lett.* **2007**, *90*, 1–3. [CrossRef]
71. Suzurikawa, J.; Nakao, M.; Jimbo, Y.; Kanzaki, R.; Takahashi, H. Light-addressed stimulation under Ca^{2+} imaging of cultured neurons. *IEEE Trans. Biomed. Eng.* **2009**, *56*, 2660–2665. [CrossRef] [PubMed]

72. Suzurikawa, J.; Kanzaki, R.; Nakao, M.; Jimbo, Y.; Takahashi, H. Optimization of thin-film configuration for light-addressable stimulation electrode. *Electron. Commun. Jpn.* **2011**, *94*, 61–68. [CrossRef]
73. Suzurikawa, J.; Nakao, M.; Jimbo, Y.; Kanzaki, R.; Takahashi, H. A light addressable electrode with a TiO_2 nanocrystalline film for localized electrical stimulation of cultured neurons. *Sens. Actuators B* **2014**, *192*, 393–398. [CrossRef]
74. Huang, Y.; Mason, A.J. Lab-on-CMOS integration of microfluidics and electrochemical sensors. *Lab Chip* **2013**, *13*, 3929–3934. [CrossRef] [PubMed]
75. Luka, G.; Ahmadi, A.; Najjaran, H.; Alocilja, E.; DeRosa, M.; Wolthers, K.; Malki, A.; Aziz, H.; Althani, A.; Hoorfar, M. Microfluidics integrated biosensors: A leading technology towards lab-on-a-chip and sensing applications. *Sensors* **2015**, *15*, 30011–30031. [CrossRef]
76. Breuer, L.; Mang, T.; Schöning, M.J.; Thoelen, R.; Wagner, T. Investigation of the spatial resolution of a laser-based stimulation process for light-addressable hydrogels with incorporated graphene oxide by means of IR thermography. *Sens. Actuators A* **2017**, *268*, 126–132. [CrossRef]
77. Miyamoto, K.i.; Ichimura, H.; Wagner, T.; Schöning, M.J.; Yoshinobu, T. Chemical imaging of the concentration profile of ion diffusion in a microfluidic channel. *Sens. Actuators B* **2013**, *189*, 240–245. [CrossRef]
78. Miyamoto, K.i.; Itabashi, A.; Wagner, T.; Schöning, M.J.; Yoshinobu, T. High-speed chemical imaging inside a microfluidic channel. *Sens. Actuators B* **2014**, *194*, 521–527. [CrossRef]
79. An, R.; Massa, K.; Wipf, D.O.; Minerick, A.R. Solution pH change in non-uniform alternating current electric fields at frequencies above the electrode charging frequency. *Biomicrofluidics* **2014**, *8*, 1–13. [CrossRef]
80. Cheng, L.J.; Chang, H.C. Microscale pH regulation by splitting water. *Biomicrofluidics* **2011**, *5*, 1–8. [CrossRef] [PubMed]
81. Cheng, L.J.; Chang, H.C. Switchable pH actuators and 3D integrated salt bridges as new strategies for reconfigurable microfluidic free-flow electrophoretic separation. *Lab Chip* **2014**, *14*, 979–987. [CrossRef] [PubMed]
82. Macounová, K.; Cabrera, C.R.; Holl, M.R.; Yager, P. Generation of natural pH gradients in microfluidic channels for use in isoelectric focusing. *Anal. Chem.* **2000**, *72*, 3745–3751. [CrossRef] [PubMed]
83. Poghossian, A.; Schöning, M. Detecting Both Physical and (Bio-)Chemical Parameters by Means of ISFET Devices. *Electroanalysis* **2004**, *16*, 1863–1872. [CrossRef]
84. Hafeman, D.G.; Harkins, J.B.; Witkowski, C.E.; Lewis, N.S.; Warmack, R.J.; Brown, G.M.; Thundat, T. Optically directed molecular transport and 3D isoelectric positioning of amphoteric biomolecules. *Proc. Natl. Acad. Sci. USA* **2006**, *103*, 6436–6441. [CrossRef]
85. Huang, S.H.; Hsueh, H.J.; Jiang, Y.L. Light-addressable electrodeposition of cell-encapsulated alginate hydrogels for a cellular microarray using a digital micromirror device. *Biomicrofluidics* **2011**, *5*, 1–10. [CrossRef]
86. Liu, Y.; Wu, C.; Lai, H.S.S.; Liu, Y.T.; Li, W.J.; Shen, Y.J. Three-dimensional calcium alginate hydrogel assembly via tiopc-based light-induced controllable electrodeposition. *Micromachines* **2017**, *8*, 1–11. [CrossRef]
87. Huang, S.H.; Wei, L.S.; Chu, H.T.; Jiang, Y.L. Light-addressed electrodeposition of enzyme-entrapped chitosan membranes for multiplexed enzyme-based bioassays using a digital micromirror device. *Sensors* **2013**, *13*, 10711–10724. [CrossRef]
88. Huang, S.H.; Wei, L.S. Light-addressed electrodeposition of polysaccharide chitosan membranes on a photoconductive substrate. *Int. J. Autom. Smart Technol.* **2014**, *4*, 196–201.
89. Andreeva, D.V.; Melnyk, I.; Baidukova, O.; Skorb, E.V. Local pH gradient initiated by light on TiO_2 for light-triggered modulation of polyhistidine-tagged proteins. *ChemElectroChem* **2016**, *3*, 1306–1310. [CrossRef]
90. Ulasevich, S.A.; Brezesinski, G.; Möhwald, H.; Fratzl, P.; Schacher, F.H.; Poznyak, S.K.; Andreeva, D.V.; Skorb, E.V. Light-induced water splitting causes high-amplitude oscillation of pH-sensitive layer-by-layer assemblies on TiO_2. *Angew. Chem. Int. Ed.* **2016**, *55*, 13001–13004. [CrossRef] [PubMed]
91. Maltanava, H.M.; Poznyak, S.K.; Andreeva, D.V.; Quevedo, M.C.; Bastos, A.C.; Tedim, J.; Ferreira, M.G.; Skorb, E.V. Light-induced proton pumping with a semiconductor: Vision for photoproton lateral separation and robust manipulation. *ACS Appl. Mater. Interfaces* **2017**, *9*, 24282–24289. [CrossRef] [PubMed]
92. Choudhury, M.H.; Ciampi, S.; Yang, Y.; Tavallaie, R.; Zhu, Y.; Zarei, L.; Gonçales, V.R.; Gooding, J.J. Connecting electrodes with light: One wire, many electrodes. *Chem. Sci.* **2015**, *6*, 6769–6776. [CrossRef] [PubMed]

93. Fabre, B. Ferrocene-terminated monolayers covalently bound to hydrogen-terminated silicon surfaces. Toward the development of charge storage and communication devices. *Acc. Chem. Res.* **2010**, *43*, 1509–1518. [CrossRef]
94. Fabre, B. Functionalization of oxide-free silicon surfaces with redox-active assemblies. *Chem. Rev.* **2016**, *116*, 4808–4849. [CrossRef]
95. Yang, Y.; Ciampi, S.; Choudhury, M.H.; Gooding, J.J. Light activated electrochemistry: Light intensity and pH dependence on electrochemical performance of anthraquinone derivatized silicon. *J. Phys. Chem. C* **2016**, *120*, 2874–2882. [CrossRef]
96. Yang, Y.; Ciampi, S.; Zhu, Y.; Gooding, J.J. Light-activated electrochemistry for the two-dimensional interrogation of electroactive regions on a monolithic surface with dramatically improved spatial resolution. *J. Phys. Chem. C* **2016**, *120*, 13032–13038. [CrossRef]
97. Lian, J.; Yang, Y.; Wang, W.; Parker, S.G.; Gonçales, V.R.; Tilley, R.D.; Gooding, J.J. Amorphous silicon on indium tin oxide: A transparent electrode for simultaneous light activated electrochemistry and optical microscopy. *Chem. Commun.* **2019**, *55*, 123–126. [CrossRef]
98. Gurrappa, I.; Binder, L. Electrodeposition of nanostructured coatings and their characterization—A review. *Sci. Technol. Adv. Mater.* **2008**, *9*, 1–11. [CrossRef]
99. Skompska, M.; Zarębska, K. Electrodeposition of ZnO nanorod arrays on transparent conducting substrates–A review. *Electrochim. Acta* **2014**, *127*, 467–488. [CrossRef]
100. Wenderich, K.; Mul, G. Methods, mechanism, and applications of photodeposition in photocatalysis: A review. *Chem. Rev.* **2016**, *116*, 14587–14619. [CrossRef] [PubMed]
101. Micheels, R.H.; Darrow, A.D.; Rauh, R.D. Photoelectrochemical deposition of microscopic metal film patterns on Si and GaAs. *Appl. Phys. Lett.* **1981**, *39*, 418–420. [CrossRef]
102. Dasog, M.; Carim, A.I.; Yalamanchili, S.; Atwater, H.A.; Lewis, N.S. Profiling photoinduced carrier generation in semiconductor microwire arrays via photoelectrochemical metal deposition. *Nano Lett.* **2016**, *16*, 5015–5021. [CrossRef]
103. Inoue, T.; Fujishima, A.; Honda, K. Photoelectrochemical imaging processes using semiconductor electrodes. *Chem. Lett.* **1978**, *7*, 1197–1200. [CrossRef]
104. Baba, R.; Konda, R.; Fujishima, A.; Honda, K. Photoelectrochemical deposition of metals on TiO_2 powders in the presence of alcohols. *Chem. Lett.* **1986**, *15*, 1307–1310. [CrossRef]
105. Lim, S.Y.; Kim, Y.R.; Ha, K.; Lee, J.K.; Lee, J.G.; Jang, W.; Lee, J.Y.; Bae, J.H.; Chung, T.D. Light-guided electrodeposition of non-noble catalyst patterns for photoelectrochemical hydrogen evolution. *Energy Environ. Sci.* **2015**, *8*, 3654–3662. [CrossRef]
106. Seo, D.; Lim, S.Y.; Lee, J.; Yun, J.; Chung, T.D. Robust and high spatial resolution light addressable electrochemistry using hematite (α-Fe_2O_3) Photoanodes. *ACS Appl. Mater. Interfaces* **2018**, *10*, 33662–33668. [CrossRef]
107. Zhong, D.K.; Cornuz, M.; Sivula, K.; Grätzel, M.; Gamelin, D.R. Photo-assisted electrodeposition of cobalt–phosphate (Co–Pi) catalyst on hematite photoanodes for solar water oxidation. *Energy Environ. Sci.* **2011**, *4*, 1759–1764. [CrossRef]
108. McDonald, K.J.; Choi, K.S. Photodeposition of Co-based oxygen evolution catalysts on α-Fe_2O_3 photoanodes. *Chem. Mater.* **2011**, *23*, 1686–1693. [CrossRef]
109. Vogel, Y.B.; Gonçales, V.R.; Al-Obaidi, L.; Gooding, J.J.; Darwish, N.; Ciampi, S. Nanocrystal inks: Photoelectrochemical printing of Cu_2O nanocrystals on silicon with 2D control on polyhedral shapes. *Adv. Funct. Mater.* **2018**, *1804791*, 1–9. [CrossRef]
110. Funt, B.L.; Tan, S.R. The photoelectrochemical initiation of polymerization of styrene. *J. Polym. Sci.: Polym. Chem. Ed.* **1984**, *22*, 605–608. [CrossRef]
111. Janáky, C.; de Tacconi, N.R.; Chanmanee, W.; Rajeshwar, K. Bringing conjugated polymers and oxide nanoarchitectures into intimate contact: Light-induced electrodeposition of polypyrrole and polyaniline on nanoporous WO_3 or TiO_2 nanotube array. *J. Phys. Chem. C* **2012**, *116*, 19145–19155. [CrossRef]
112. Okano, M.; Itoh, K.; Fujishima, A. Incorporation of dye molecules in polypyrrole films in the process of photoelectrochemical deposition. *Chem. Lett.* **1987**, *16*, 2129–2130. [CrossRef]
113. Okano, M.; Kikuchi, E.; Itoh, K.; Fujishima, A. Photoelectrochemical polymerization of pyrrole on ZnO and its application to conducting pattern generation. *J. Electrochem. Soc.* **1988**, *135*, 1641–1645. [CrossRef]

114. Okano, M.; Baba, R.; Itoh, K.; Fujishima, A. New aspects in area-selective electrode reactions on illuminated semiconductors. In *Studies in Surface Science and Catalysis*; Elsevier: Amsterdam, The Netherlands, 1989; Volume 47, pp. 375–387.
115. Takagi, S.; Makuta, S.; Veamatahau, A.; Otsuka, Y.; Tachibana, Y. Organic/inorganic hybrid electrochromic devices based on photoelectrochemically formed polypyrrole/TiO_2 nanohybrid films. *J. Mater. Chem.* **2012**, *22*, 22181–22189. [CrossRef]
116. Welden, R.; Scheja, S.; Schöning, M.J.; Wagner, P.; Wagner, T. Electrochemical Evaluation of Light-Addressable Electrodes Based on TiO_2 for the Integration in Lab-on-Chip Systems. *Phys. Status Solidi (a)* **2018**, *215*, 1800150. [CrossRef]
117. Janáky, C.; Chanmanee, W.; Rajeshwar, K. Mechanistic aspects of photoelectrochemical polymerization of polypyrrole on a TiO_2 nanotube array. *Electrochim. Acta* **2014**, *122*, 303–309. [CrossRef]
118. Kierstan, M.; Bucke, C. The immobilization of microbial cells, subcellular organelles, and enzymes in calcium alginate gels. *Biotechnol. Bioeng.* **1977**, *19*, 387–397. [CrossRef]
119. Gurtner, C.; Edman, C.F.; Formosa, R.E.; Heller, M.J. Photoelectrophoretic transport and hybridization of DNA oligonucleotides on unpatterned silicon substrates. *J. Am. Chem. Soc.* **2000**, *122*, 8589–8594. [CrossRef]
120. Chow, B.Y.; Emig, C.J.; Jacobson, J.M. Photoelectrochemical synthesis of DNA microarrays. *Proc. Natl. Acad. Sci. USA* **2009**, *106*, 15219–15224. [CrossRef]
121. Jacobs, J.W.; Rikken, J.M. Photoelectrochemically-induced gold deposition on p-GaAs electrodes part I. Nucleation and growth. *J. Electrochem. Soc.* **1989**, *136*, 3633–3640. [CrossRef]
122. Ngaboyamahina, E.; Cachet, H.; Pailleret, A.; Sutter, E. Photo-assisted electrodeposition of an electrochemically active polypyrrole layer on anatase type titanium dioxide nanotube arrays. *Electrochim. Acta* **2014**, *129*, 211–221. [CrossRef]

Review

Recent Developments of High-Resolution Chemical Imaging Systems Based on Light-Addressable Potentiometric Sensors (LAPSs)

Tao Liang [1,2], Yong Qiu [1], Ying Gan [1], Jiadi Sun [1], Shuqi Zhou [1], Hao Wan [1,2,*] and Ping Wang [1,2,*]

1 Biosensor National Special Laboratory, Key Laboratory of Biomedical Engineering of Ministry of Education, Department of Biomedical Engineering, Zhejiang University, Hangzhou 310027, China; cooltao@zju.edu.cn (T.L.); zjubme_qy@zju.edu.cn (Y.Q.); ganying@zju.edu.cn (Y.G.); 3130102231@zju.edu.cn (J.S.); deardme@zju.edu.cn (S.Z.)
2 State Key Laboratory of Transducer Technology, Shanghai 200050, China
* Correspondence: wh1816@zju.edu.cn (H.W.); cnpwang@zju.edu.cn (P.W.)

Received: 31 August 2019; Accepted: 1 October 2019; Published: 3 October 2019

Abstract: A light-addressable potentiometric sensor (LAPS) is a semiconductor electrochemical sensor based on the field-effect which detects the variation of the Nernst potential on the sensor surface, and the measurement area is defined by illumination. Thanks to its light-addressability feature, an LAPS-based chemical imaging sensor system can be developed, which can visualize the two-dimensional distribution of chemical species on the sensor surface. This sensor system has been used for the analysis of reactions and diffusions in various biochemical samples. In this review, the LAPS system set-up, including the sensor construction, sensing and substrate materials, modulated light and various measurement modes of the sensor systems are described. The recently developed technologies and the affecting factors, especially regarding the spatial resolution and temporal resolution are discussed and summarized, and the advantages and limitations of these technologies are illustrated. Finally, the further applications of LAPS-based chemical imaging sensors are discussed, where the combination with microfluidic devices is promising.

Keywords: LAPS; chemical imaging; spatial and temporal resolution; semiconductor; microfluidics

---

1. Introduction

The light-addressable potentiometric sensor (LAPS) [1–6] is a spatially resolved biochemical sensor based on the field-effect with an electrolyte-insulator-semiconductor (EIS) structure [7–11], in which the sensing surface of the insulating layer is in contact with the analyte solution. It shares the same structure as ion-selective field-effect transistors (ISFETs) and EIS capacitive sensors. The ISFET detects variation of the source–drain conductance, while the EIS capacitive sensor and LAPS detect the variation of the capacitance of the depletion layer, and the thickness of which responds to the potential of the sensing surface, thereby obtaining the analyte concentration. The characteristics of these field-effect devices have been compared and summarized systematically [12–15]. LAPS is developed from the combination of scanned light pulse technique (SLPT) method and EIS structures [6]. Figure 1a displays the schematic of a typical LAPS set-up. A standard set-up usually applies a three-electrode system, including a reference electrode (RE) such as Ag/AgCl, a counter electrode such as platinum wire, and a semiconductor substrate (silicon bulk) as a working electrode (WE). An ohmic contact is formed on the back of the semiconductor substrate to connect with the WE. The depletion layer forms at the semiconductor/insulator interface when a DC bias voltage is applied between the electrolyte and the semiconductor. After illuminating a frequency-modulated light at the top or bottom of the sensor, hole–electron pairs are generated. The n these photocarriers will be separated in the electric field and generate AC photocurrent, although some of them may recombine. This AC photocurrent has the same frequency as the modulated light and can be detected by the peripheral circuit.

LAPS sensors record local signals with spatial resolution because the photocurrent is generated only in the illuminated area, thereby achieving chemical imaging with a scanning light spot. This feature can not only measure local surface potential changes, but also the local impedance. The latter application is termed scanning photo-induced impedance microscopy (SPIM) [16–18]. The photocurrent-voltage (I–V) curves are shown in Figure 1b, LAPS detects the surface potential changes by recording the shift in the depletion region of the I-V curve along the voltage axis (shift of curve A to B), while SPIM measures the impedance changes by detecting the change of saturation current value in the inversion region (shift of curve A to C).

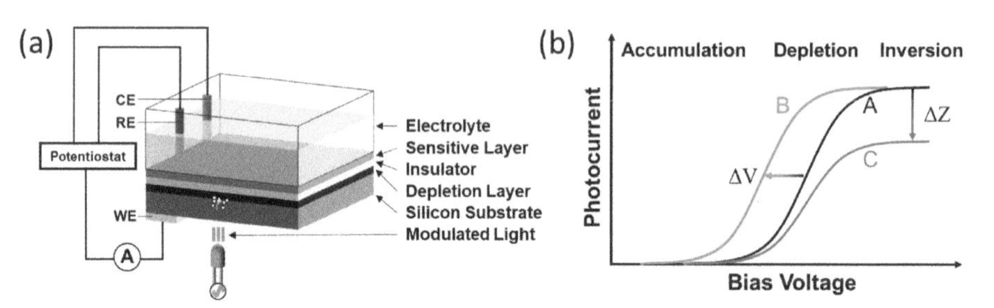

Figure 1. (**a**) Schematic of a typical LAPS set-up; (**b**) I-V curves representing the potential shift (LAPS) and impedance changes (SPIM).

Since the first report in 1988 by Hafeman et al. [1], LAPS has been developed widely for sensing and imaging of (bio)chemical species [19–24] and living cells [25,26]. Such type of LAPS was initially termed pH-imaging sensor, as it first achieved the visualization of pH distribution [27–33]. After that, chemical imaging sensor was introduced to refer to the sensors that visualize the distribution of other chemical species, such as the enzyme [34], microcapsules [35] and ion diffusion [36,37].

The LAPS chemical imaging sensor has the following features and advantages. In contrast to the conventional electrode/sensor array, LAPS has no multi-pixel structure, and the pixel layout is defined by the scanning light, so the imaging region can be flexibly adjusted according to demand. Without the limitation of the pixel number, the image resolution is mainly determined by light spot size and scanning step distance. Besides, no additional wires are required, regardless of the imaging area and resolution which simplifies the structure of the circuit, and makes it durable for long-term use. Finally, LAPS can be integrated easily with microfluidic devices due to the flatness and uniformity of the sensing surface, which greatly increases the potential of LAPS for biomedical and micro liquid detection applications.

In recent years, various aspects of LAPS applications, including chemical species detection and surface modification have been comprehensively reviewed. Herein, we discuss and focus on the most advanced technologies of high-resolution imaging in the spatial and temporal aspects of LAPS-based chemical imaging sensor systems. The sensor system set-up, including the sensor chip construction, materials for sensing layer and substrate, modulated light property and various measurement modes are described and summarized. The prospects of future applications combined with microfluidic devices are discussed.

2. Measurement System Set-Up

The LAPS system for chemical imaging consists of a LAPS chip, a modulated light source, a mechanical device to move sensor plate or the position of the light spot to achieve the scanning.

It also includes peripheral measurement circuit, electrochemical set-up like the three-electrode system, and PC software. The commonly used set-ups are discussed separately below.

2.1. Sensor Construction and Sensing Materials

A typical LAPS chip has a sensing layer/insulator/semiconductor substrate structure. The semiconductor is usually made of silicon bulk, covered with a thermally oxidized SiO_2 layer. In order to reduce the recombination of photocarriers and improve spatial resolution, the silicon bulk can be replaced with thin silicon films (thinned silicon substrate [38–40], silicon on insulator (SOI) [18], amorphous silicon (a-Si) [41–44] and silicon on sapphire (SOS) [45,46]) or other semiconductor thin layers deposited on transparent substrates (GaAs [47], GaN [48], In-Ga-Zn oxide (IGZO) [49] and indium tin oxide (ITO) [50]). The sensing material such as Si_3N_4 [1,28,38,39,51], Al_2O_3 [52–54], Ta_2O_5 [51], HfO_2 [55–57], NbO_x [40,58,59], SnO_x [60], self-assembled organic monolayer (SAM) [61,62] and nanofibers [63,64] can be used for pH sensing. Other chemical species can be also detected by modifying the sensing surface with appropriate materials. The deposition of chalcogenide glass films [65–69] can detect metal ions such as Pb^{2+}, Fe^{3+}, Cd^{2+} and Cr^{6+}. The detection of inorganic salt ions (e.g., K^+, Ca^{2+}, Na^+, Mg^{2+}, Zn^{2+}, Li^+, NO_3^-, CO_3^{2-}, and SO_4^{2-}) can be achieved by depositing polymer membranes carrying ionophores such as PVC membranes [53,70–76], silicone rubber membranes [77–81] and photocurable membranes [82]. Immobilization of enzymes [83–90], antibodies [23,91,92] or DNA [93–96] on the sensing layer enables the detection of biomolecules such as glucose, urea, penicillin G, antigens, immunoglobulins G (IgG) and target DNA, etc. The sensing materials and target species have been systematically summarized [20–22].

2.2. Modulated Light

The modulated light illuminating the LAPS chip is usually sinusoidally modulated with a frequency in the range of 50 Hz to 100 kHz [97–99]. It is necessary to use the modulated light with a photon energy greater than the band gap. For silicon-based chips, visible or near-infrared light is commonly used. In terms of spatial resolution, according to the Rayleigh criterion, the resolution can be improved by reducing the light wavelength [45,61]. Regarding the illumination direction of the modulated light, there are two kinds of settings: Frontside and backside illuminations [97,100,101]. The backside illumination requires photocarriers to diffuse across the semiconductor substrate, which causes more recombination of the carriers and weakening the photocurrent signal. The long lateral diffusion distance will also reduce the spatial resolution. The frontside illumination directly generates photocarriers in depletion layer, and it has the advantages of larger signal amplitude and higher spatial resolution. However, the light needs to pass through the solution, specimen or cells, the scattering and absorption will disturb the light uniformity [102]. The refore, the backside illumination is more practical and often applied.

2.3. Measurement Modes

LAPS is a kind of potentiometric sensor based on the field-effect, in which the sensing surface potential responds to the analyte concentration. The specific adsorption of the sensitive layer to the analyte changes the sensor surface potential. This sensing mechanism is essentially the same as ISFET and has already been explained systematically [9,103–106]. When the surface potential changes in response to the analyte concentration, the I-V curve also shifts accordingly. This change can be quantified by measuring the horizontal shift of the inflection point (that is calculating the zero-crossing point in the second derivative of the curve [3,28]). The n we can obtain the calibration curve between the inflection point voltage and the analyte concentration to achieve the measurement purpose. In most applications, we only need the changes of bias voltage at the inflection point. Acquiring the complete I–V curves at all pixels is too time consuming, especially in imaging applications. In order to meet different needs, the following measurement modes are usually used in different applications. The ir pros and cons are summarized in Table 1.

2.3.1. Constant-Voltage Mode

The method of obtaining the I-V curve of LAPS is to apply a scan bias voltage and then record the change in photocurrent with voltage. If there is no need to get a complete curve, we can just fix a constant bias and then measure the corresponding photocurrent. This constant-voltage mode (also called constant-bias mode) is fast and suited for multi-pixel imaging, thus it is usually used in the chemical imaging applications. As the surface potential of the sensor changes with the analyte concentration, the obtained photocurrent can be converted into the bias voltage by assuming a linear slope of the transition part of the I-V curve [27]. The voltage near the inflection point is usually chosen as the fixed bias, where the linearity and sensitivity is relatively good [107]. The constant-voltage mode has been used to obtain the pH image of *Saccharomyces cerevisiae* colonies [28] and to investigate the spatial resolution of LAPS [101]. However, the photocurrent variation of the constant-voltage mode is limited inside the transition part of the I–V curve. If the analyte concentration varies too large, the photocurrent will be saturated. Other than that, the assumed slope may cause some potential errors [107].

2.3.2. Constant-Current Mode

Different from the constant-voltage mode, the constant-current mode chooses a constant current and then recording the change of the applied bias voltage, which requires a feedback loop adjusting the bias to maintain the photocurrent value in constant [4,107]. The results of this mode are more accurate because the recorded bias voltage change directly reflects the change of the analyte concentration, and does not need to be converted by the assumed slope, thereby avoiding some potential errors. In addition, there is no limit to the detection range, which allows the constant-current mode to measure larger analyte concentration variation. One constant current value can be set for all pixels of the ideal imaging sensor. However, general cases usually require two scans [21] since the chemical imaging sensor is not spatially uniform. The first scan records the initial photocurrent values of all pixels, and the second scan reproduces the photocurrent at each pixel. In this mode, an additional time is required until the sensor capacitance is charged when the bias voltage is updated. The refore, the constant-current mode is more time consuming [38] and it is often used in analyte sensing applications rather than chemical imaging.

2.3.3. Potential-Tracking Mode

In order to improve the accuracy of the results without sacrificing measurement time, Miyamoto et al. [27] proposed a new data acquisition method, namely the potential-tracking mode. In this mode, dozens of bias voltages are selected and the corresponding photocurrent values are recorded at each pixel. The n the entire I-V curve can be reconstructed by curve-fitting. Compared to measuring the complete I-V curve at each pixel, this mode can also obtain the shift of the entire I-V curve with shorter measurement time. Contrast with the constant-voltage mode, the proposed potential-tracking mode is able to measure a larger variation of analyte concentration, and the shift of I-V curve can more accurately reflect the variation of the analyte concentration. However, the potential-tracking mode also requires additional charging time to accommodate the new bias voltage, and requires an additional step of curve-fitting during the data processing.

2.3.4. Phase Mode

For LAPS, the semiconductor substrate absorbs photon energy to generate hole-electron pairs, thereby the amplitude of sensor signal will be significantly affected by fluctuations in light intensity and the defects of semiconductor substrate [33,108]. In order to achieve accurate measurement, the phase-mode [109] was proposed to eliminate these effects. This mode detects the phase variation of the photocurrent, instead of the amplitude as in the common measurement mode. The AC photocurrent and the modulation signal are simultaneously recorded and then the phase difference between them are calculated. The phase-voltage curve shifts along the voltage axis in response to the analyte concentration variation, similar to the conventional I-V curve. The phase-mode is much less sensitive to the loss of photocarriers (caused by light intensity fluctuation and semiconductor defects), which contributes to the improvement of chemical image uniformity. During chemical imaging, the phase variations are recorded under a constant bias voltage. Errors may be also caused by the assumed slope during the phase-voltage conversion.

2.3.5. Pulse-Driven Mode

The pulse-driven mode [110] utilizes a pulse-modulated light to generate the photocurrent rather than a conventional continuously modulated light. The photocurrent is mainly generated by two parts: photocarriers generated in the depletion layer and the diffused photocarriers. The utilization of a short light pulse with a high intensity can effectively eliminate the influence of the diffused photocarriers, and the shorter integration time of light pulse contributes to the improvements of spatial resolution (by a factor of 6 or more) and the contrast of line scan (by a factor of 3 or more), but it also results in a lower signal-to-noise ratio (SNR). In the pulse-driven mode, the photocurrent signal is acquired using a charge amplifier based on a high-speed operational amplifier rather than a trans-impedance amplifier to follow the fast change resulting from a light pulse. The measurement in pulse-driven mode takes more time than continuous modulation modes, because it needs extra time intervals to release the carrier distribution within the semiconductor between single pulses.

Table 1. Pros and cons of LAPS measurement modes.

Measurement Modes	Pros	Cons	Reference
Constant-Voltage Mode	Rapid Measurement; Suitable for Multi-Pixel Imaging	Small Detection Range; Potential Conversion Rrrors	[28,101]
Constant-Current Mode	Unlimited Detection Range; Accurate Measurement	Necessary Feedback Loop; Long Measurement Time	[4,107]
Potential-Tracking Mode	Entire I-V Curves; Unlimited Detection Range; Relatively Accurate	Additional Curve-Fitting; Necessary Charging Time	[27]
Phase Mode	Good Robustness; Good Imaging Uniformity	Necessary Simultaneous Record of Photocurrent and Modulation Signal; Potential Conversion Errors	[109]
Pulse-Driven Mode	High Spatial Resolution; High Contrast of Line Scan	Low SNR; Long Measurement Time	[110]

3. Spatial Resolution

Spatial resolution is one of the most important features of a chemical imaging LAPS; it has been defined as the smallest size of the pattern or structure that can be resolved by LAPS [111]. Photoresist patterns are deposited onto the insulator and then the resolution can be determined as the distance for achieving a photocurrent drop when scanning from an area without photoresist to another area with photoresist, by calculating the full width at half-maximum (FWHM) of the first derivative of photocurrent response across the edge of photoresist pattern [45,61]. A metal gate has also been used to determine the spatial resolution by measuring the photocurrent decay outside the gate area [17,41].

As a LAPS chip has no isolation between neighboring pixels, the spatial resolution is mainly determined by the lateral diffusion of minority photocarriers out of the illuminated area in the semiconductor layer [19,97,112]. Improving the spatial resolution can be carried out in two aspects: semiconductor substrate and modulated light property.

3.1. Semiconductor Substrate

The oretical calculations showed that the resolution is determined by the lateral diffusion length of photocarriers in the semiconductor substrate [99,100,102]. In single crystalline silicon substrate, the diffusion length can be hundreds of micrometers. High doping concentration will increase the recombination of photocarriers and reduce the diffusion length, obtaining a higher spatial resolution [100,101,113]. However, more recombination of photocarriers also reduces the photocurrent signal, resulting in a loss of sensitivity.

Another approach to improving the spatial resolution is using the semiconductor material with a shorter diffusion length, such as GaAs [47], GaN [48] and amorphous silicon (a-Si) [41–44]. The shorter diffusion length also means more loss of photocarriers, resulting in a smaller photocurrent signal [97], which requires the semiconductor substrate must be thin. A diffusion length of 3.1 μm was measured using an 8 μm epilayer of GaAs [47], and a resolution of a thin film a-Si deposited on a transparent glass substrate is in the submicron range [41] with a diffusion length of 120 nm [114].

The simplest and most effective method to improve the spatial resolution is reducing the thickness of the semiconductor substrate [29,31,97,101,111,112]. With a thicker substrate, the photocarriers will travel a longer distance to arrive the depletion layer, causing a larger lateral diffusion length and a decreased resolution. Nakao et al. thinned the silicon substrate from 300 to 100 μm and further to 20 μm, achieved spatial resolutions of 500, 200 and 10 μm [31]. However, a silicon substrate thinner than 100 μm is fragile, and cannot provide a good mechanical support for practical applications. This problem can be solved by using the device with a thin semiconductor layer such as silicon on insulator (SOI) or silicon on sapphire (SOS). The use of SOI with a 7 μm thick device layer obtained an effective photocarriers diffusion length of about 13 μm and the diffusion length of SOS with 1 μm thick silicon layer was 570 nm [17]. Photocarriers are generated only in the light spot area, not the entire semiconductor layer, resulting in the elimination of the effects caused by stray light and reflections at edges. A resolution of 5 μm was obtained with another SOS device with 0.5 μm thick silicon layer [111].

A transparent substrate such as sapphire can provide good mechanical support for a thin semiconductor layer, and the excellent light transmittance can achieve high spatial resolution in the case of backside illumination. Since it is difficult to grow a conventional thermal oxide layer on SOS [20], the insulating layer can be replaced with a good anodic oxide [45] or a self-assembled monolayer [61]. Currently, the best spatial resolution of LAPS/scanning photo-induced impedance microscopy (SPIM) with backside illumination is 0.8 μm, achieved using a SOS substrate with a 0.5 μm thick silicon layer and an anodic oxide [45]. An SU-8 pattern was fabricated on the insulator and the photocurrent image is shown in Figure 2a. Modulated lights with different wavelengths of 405, 633, 1064 and 1250 nm were used and obtained the spatial resolutions of 1.5, 2.2, 3.0 and 0.8 μm respectively. The normalized photocurrents curves scanning across the edge of the photoresist at these wavelengths are displayed in Figure 2b.

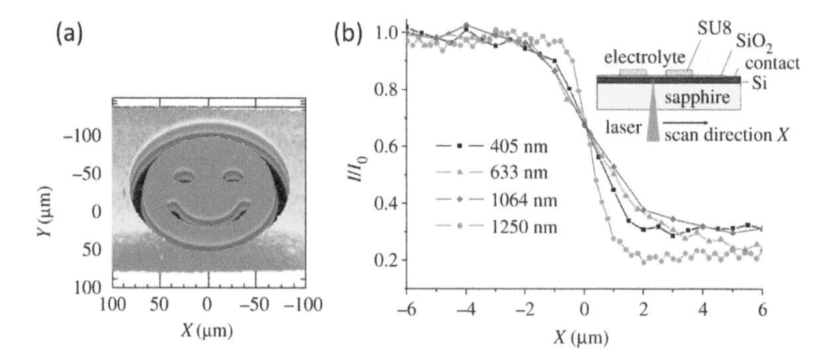

Figure 2. (**a**) Photocurrent image of the SU-8 pattern; (**b**) Normalized photocurrent curves scanning across the edge of the photoresist at different wavelengths. Reprinted (adapted) with permission from Chen et al. [45] Copyright (2019) Elsevier.

Although SOS has many advantages, the high cost limited its widespread application. In recent years, indium tin oxide (ITO) coated glass has been chosen as a new substrate for LAPS and SPIM as a bioelectronic taste sensor for bitter [115] or a high resolution imaging sensor [50] due to the features of low cost, robustness, stability and easy to be modified. The LAPS with ITO coated glass as the substrate has electrolyte–semiconductor (ES) structure without an insulator. The schematic of the LAPS set-up is shown in Figure 3a. The wavelength of the modulated light illuminated from back side is 405 nm. The verification of UV-vis spectrum and impedance spectrum proved that the excitation light wavelength is in the UV part and the blocking nature of the ITO-solution interface in the dark. Different from the conventional structure, the LAPS with ITO coated glass has no insulating layer, and the measured photocurrent signal is generated from the redox currents, which is affected by the pH dependent kinetics in the anodic oxidation process. The average pH sensitivity of this ITO coated glass LAPS is about 70 mV/pH, higher than the Nernstian theoretical value of 59 mV/pH. Although high sensitivity is good for sensors, this super Nernstian sensitivity is usually caused by multiple potential changes, and the selectivity of the ITO coated glass sensor has not been verified. Poly (methyl methacrylate) (PMMA) dots on the ITO surface wasre used to characterize the lateral resolution. Figure 3b shows the photocurrent image of the dot pattern. Due to the high impedance of PMMA, the photocurrent in polymer dot area decreased to nearly zero. To determine the lateral resolution, a line scanning across the edge of the dot was carried out at a step of 1 μm. The normalized photocurrent-position curve is shown in Figure 3c. The full width at half-maximum (FWHM) of the first derivative of photocurrent response across the edge is usually used to indicate the lateral resolution. As shown in Figure 3d, the FWHM (lateral resolution) is about 2.3 μm, close to the 1.5 μm resolution of the SOS at the wavelength of 405 nm. This indicates that the ITO coated glass can be a low-cost alternative material for SOS for high resolution imaging.

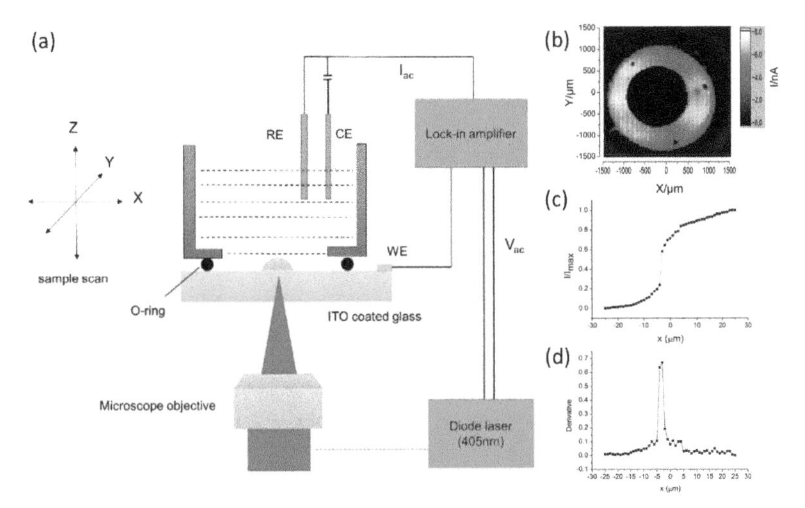

Figure 3. (**a**) Schematic of the LAPS set-up with ITO coated glass; (**b**) The LAPS image of the PMMA dot; (**c**) Normalized photocurrent-position curve scanning across the edge of PMMA dot; (**d**) the first derivative of (B) with the FWHM of 2.3 μm. Reprinted (adapted) with permission from Zhang et al. [50] Copyright (2019) American Chemical Society.

Recently, the LAPS with ITO coated glass substrate has been developed into a novel photoelectrochemical imaging system (PEIS) for single live cells imaging [116]. The mapping of surface charge of the cell bottom can be obtained with the set-up shown in Figure 4a, the photocurrents in this structure features the redox currents. The photocurrent image of MG63 human osteosarcoma cells by PEIS is shown in Figure 4b, while the optical image by CMOS camera is shown in Figure 4c, indicates a great correlation with the former. The photocurrent in the cell area was smaller, this may be caused by the negative charge of the cell surface. In addition, the process of cell lysis can be also monitored with PEIS. Human mesenchymal stem cells (hMSC) and 0.04% *v/v* TX-100 (0.68 mM) in Dulbecco's Phosphate Buffered Saline (DPBS) was used for cell membrane permeabilization and cell lysis. Initially, the PEIS image (Figure 4d) and optical image (Figure 4e) without TX-100 shows a clear photocurrent distribution and the intact profile of cell membrane. 30 min after the TX-100 addition (Figure 4f,g), the cell has lost its intact form, the increased photocurrent in the cell area indicating the progressive solubilization of cell membrane. And after 200 min, the cell profile disappeared in the PEIS image (Figure 4h), indicating that the cell membrane has completely broken down. This is consistent with the optical image (Figure 4i) where only the residues of the cell remained. LAPS has already achieved the visualization of the defects recovery process in a cultured cell layer [25], now with the help of the abovementioned PEIS, the visualization extent of the metabolism and lysis can be accurate to the single cell level, which expands the application prospects of LAPS imaging.

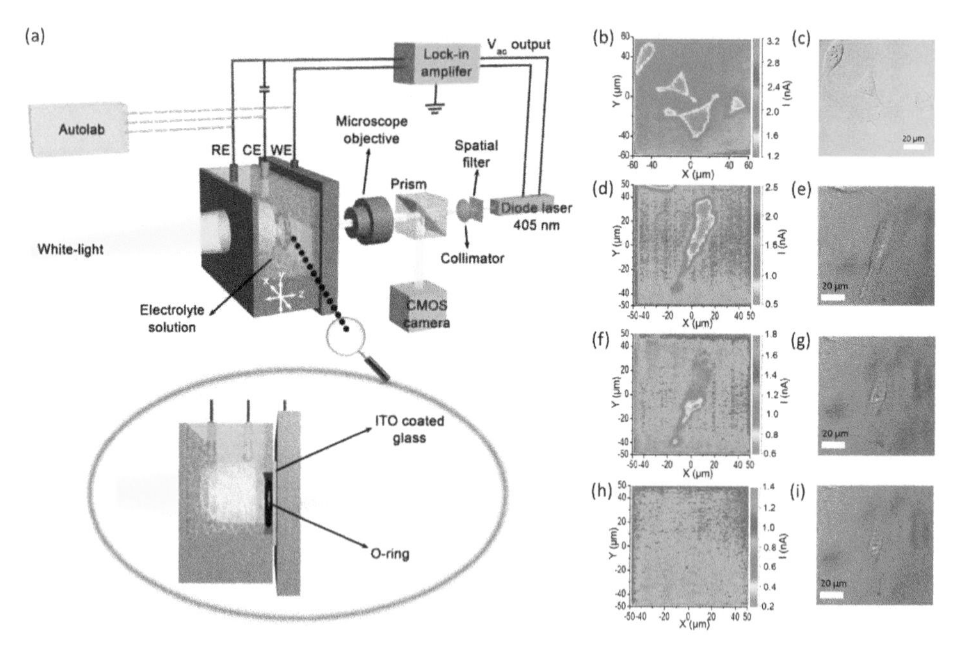

Figure 4. (**a**) Schematic of the PEIS set-up with ITO coated glass; (**b**) photocurrent image by PEIS and (**c**) optical image by camera of MG63 human osteosarcoma cells; (**d**) PEIS image and (**e**) optical image of hMSC before TX-100 addition; (**f**) PEIS image and (**g**) optical image after 30 min incubation with 0.04% TX-100 in DPBS; (**h**) PEIS image and (**i**) optical image after 200 min incubation with 0.04% TX-100 in DPBS. Reprinted from Wu et al. [116].

3.2. *Property of Modulated Light*

In addition to the choice of semiconductor substrate, the modulated light property also affects the spatial resolution. Reducing the light spot size is a direct way to increase the spatial resolution. Nakao et al. [28] focused the modulated light spot to 1 μm size for the first time and obtained the pH distribution image of the *Saccharomyces cerevisiae* colonies. In previous reports, the effect of the light wavelength on the resolution has been investigated [31,45,61,100,113].

The use of infrared light can improve the spatial resolution. For the bulk silicon substrates with backside illumination, the light with long wavelength can reach deeper into the silicon and photocarriers will be generated nearer to the depletion layer. 10 μm test patterns can be resolved with an infrared light (the wavelength is 830 nm) illuminating a 20 μm silicon substrate [31], in which the penetration depth is 13 μm.

To improve spatial resolution, an auxiliary illumination of constant light can be added around the modulated light [117–120]. As shown in Figure 5a, a ring-shaped constant illumination increases the photocarriers concentration, and the enhanced recombination will block the lateral diffusion of photocarriers induced by the modulated light. The effect was predicted by simulation [117], and the experiment [119] verified a resolution improvement from 92 μm without to 68 μm with the ring-shaped constant illumination by bundled fiber-optics (Figure 5b). Simulations [118] suggest that the resolution would be limited due to the enhanced recombination, causing a decrease of the photocurrent amplitude. And the spread of light after exiting the end of fiber-optics will blur the shapes of both the modulated light spot and the ring-shaped constant illumination. In order to further improve the resolution, the hybrid illumination system based on a binocular tube can be used instead of the bundled fiber-optics [120]. The binocular tube optics set-up is shown in Figure 5c and it can resolve a 100 μm pattern while the bundled fiber-optics cannot, indicating that the spatial resolution has been further improved.

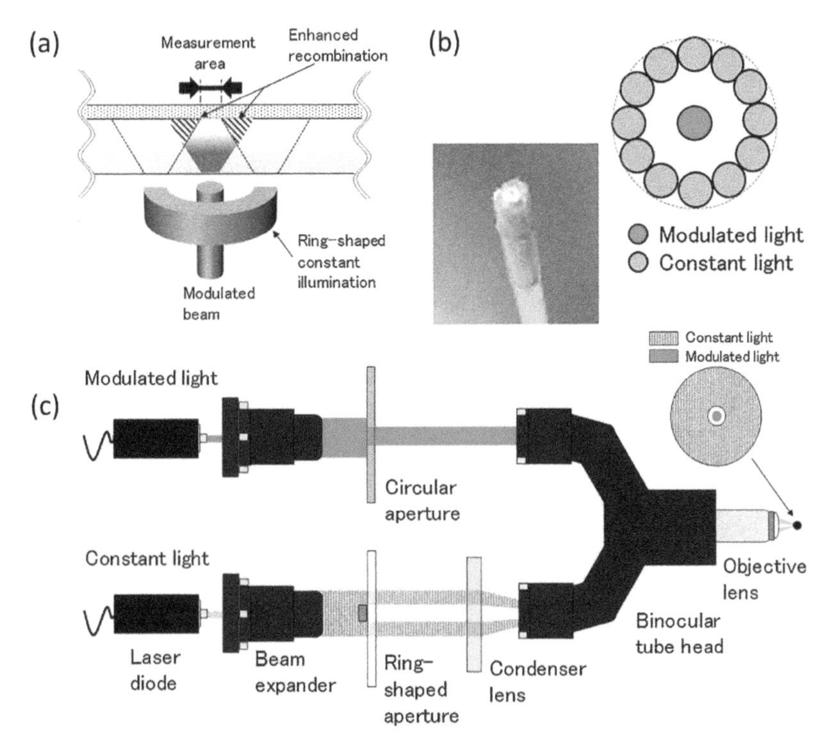

Figure 5. (**a**) Schematic of the LAPS with hybrid illumination; (**b**) Assembly of the ring-shaped bundled fiber-optics; (**c**) Schematic of the hybrid illumination set-up based on binocular tube optics. Reprinted (adapted) with permission from Miyamoto et al. [120] Copyright (2019) Elsevier.

The use of pulse-driven light [110] can also improve the spatial resolution, which has been discussed in Section 2.3.5 of this article. For the thin silicon substrates such as SOS [45,61], the penetration depth of the light is less important. The effect of wavelength on spatial resolution is mainly determined by the Rayleigh criterion: $r = 0.61\lambda/NA$, where r is the resolution, λ is the wavelength and NA is the numerical aperture of the microscope objective. In the wavelength range of 405 nm to 1064 nm, the spatial resolution increases as the wavelength decreases. The best resolution is 1.5 μm at the wavelength of 405 nm. Further improvement of the resolution to 0.8 μm can be achieved based on the two-photon effect with a 1250 nm light. In this case, the photon energy of the modulated light was smaller than the energy bandgap of silicon, the photocarriers were generated under the two-photon effect only near the focused spot close to the depletion layer [17]. All the technical information about spatial resolution mentioned above is summarized in Table 2.

Table 2. Technical information about improving the spatial resolution of LAPS.

Methods		Notes	Sensor Construction	Modulated Light Parameters [1]	Spatial Resolution	Reference
Semiconductor Substrate Properties	High Doping Concentration	Simulation Results	50 nm SiO_2/50 nm Si_3N_4/200 μm Si	Backside Illumination; λ = 800 nm, f = 5 kHz, P = 6 W/cm^2, S = 20 μm	<30 μm	[100]
	Short Diffusion Length Materials	Semiconductor Material: GaAs	8 μm Pt/100 nm Anodic Oxide/8 μm Epilayer of GaAs	Frontside Illumination; λ = 780 nm, f = 10 kHz, P = 0.18 mW, S = 2.6 μm	3.1 μm	[47]
		Semiconductor Material: Amorphous Si	20~150 nm Metal Gate/50 nm Si_3N_4/30 nm SiO_2/0.3~1.5 μm a-Si/Glass/500 nm Al/700 ZnO	Frontside Illumination; λ = 430 nm, P = 1 mW, S = 1.03 μm	<1 μm	[41]
	Thinned Semiconductor Substrate	Infrared Light	100 nm Au/Photoresist Pattern/100 nm Si_3N_4/50 nm SiO_2/20 μm Si	Backside Illumination; λ = 830 nm, f = 1~10 kHz, S = ~1 μm	<10 μm	[31]
	Transparent Substrate with Thin Semiconductor Layer	Silicon on Sapphire	Photoresist Pattern/6.7 nm Anodic Oxide/0.5 μm Si/500 μm Sapphire/20 nm Cr/80 nm Au	Backside Illumination; λ = 405 nm, f = 1 kHz, P = 1 mW (Single Photon Effect)	1.5 μm	[45,61]
		ITO Coated Glass; No Insulator	PMMA dot/~140 nm ITO/500 μm Glass	Backside Illumination; λ = 405 nm, f = 10 Hz, S = ~1 μm	2.3 μm	[50,116]
Modulated Light Properties	Small Light Spot Size	Spot Size at Micron Level	100 nm Si_3N_4/50 nm SiO_2/300 μm Si/AuSb	Backside Illumination; λ = 633 nm, f = 1~10 kHz, P = 10 mW, S = ~1 μm	<500 μm	[28]
	Infrared Light	Thin Silicon Substrate	100 nm Au/Photoresist Pattern/100 nm Si_3N_4/50 nm SiO_2/20 μm Si	Backside Illumination; λ = 830 nm, f = 1~10 kHz, S = ~1 μm	< 10 μm	[31]
	Auxiliary Illumination	Ring-Shaped Constant Light	50 nm Si_3N_4/50 nm SiO_2/200 μm Si/Au	Backside Illumination; λ = 832 nm, Modulated, P = 0.002 mW; λ = 832 nm, Constant, P = 0.1 mW	<68 μm	[119,120]
	Pulse-Driven Modulated Light	Charge Amplifier	40 nm Si_3N_4/40 nm SiO_2/200 μm Si/Au	Backside Illumination; λ = 905 nm, t = 2 μs, P = 85 mW, S = ~1.1 μm	110 μm	[110]
	Two-Photon Effect	Silicon on Sapphire	Photoresist Pattern/6.7 nm Anodic Oxide/0.5 μm Si/500 μm Sapphire/20 nm Cr/80 nm Au	Backside Illumination; λ = 405 nm, f = 1 kHz, P = 1 mW (Two-Photon Effect)	0.8 μm	[45,61]

[1] λ: Wavelength; f: Frequency; P: Power; t: Integration time; S: Light spot size.

4. Temporal Resolution

A typical way of LAPS imaging is to use a single beam scan to readout the photocurrent values of each pixel one by one. The image acquisition time is given by multiplying the measurement time of each pixel by the number of pixels. In a typical set-up, the measurement time per pixel is millisecond level, and the acquisition time for one image is on the order of minutes at a resolution of 128 × 128 pixels, for example. In addition, the scanning speed of the mechanical stage also greatly limits the temporal resolution. The minute-level time consuming is acceptable for monitoring some slow processes, such as metal corrosion [121] and cell growth [25,30], but difficult for the fast processes such as chemical reactions and ion diffusion. The refore, it is necessary to reduce the measurement time.

4.1. Single Modulated Light Without Mechanical Movement

An analog micromirror based on micro-electro-mechanical systems (MEMS) has been used to replace the traditional mechanical stage [122–124]. The schematic of this set-up is shown in Figure 6a, the light spot can be moved and located precisely on the back of the sensor plate according to user requirements by PC software control. This set-up was able to acquire a chemical image with the resolution of 500 × 400 pixels within 40 s, and can also record the spatiotemporal change in pH distribution in KCl solution and buffer solution with a rate of 16 frames per second (fps) and a resolution of 10 × 8 pixels per frame. However, in this way, the angle of light illumination at each pixel is different, resulting in different spot sizes on the back of the sensor plate, which affects the uniformity of the image.

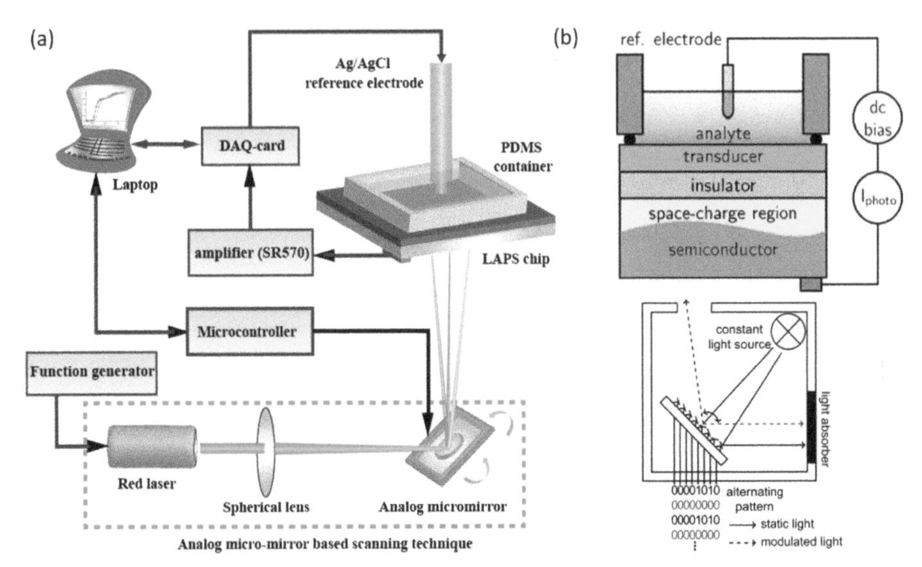

Figure 6. (**a**) Schematic of the LAPS imaging system based on analog micromirror. Reprinted from; (**b**) Schematic of the LAPS imaging system based on the DLP, the light is modulated by the fast switching mirrors and the position and size of the light spot can be adjusted as required. Reprinted (adapted) with permission from Das et al. [124] and Wagner et al. [125] Copyright (2019) Elsevier.

In order to completely avoid the effects of mechanical motion (including the rotation of the micromirror), digital micromirror device (DMD) based on the digital light processing (DLP) technology was implemented as the scanning light source for LAPS [125,126]. The system set-up is shown in Figure 6b. The modulated light is generated by the fast switching DMD, the size and shape of light spot can be easily defined according requirements without any modification of hardware. This set-up has 480 × 320 micromirrors, resulting in a maximum resolution of 153,600. The authors investigated a method that uses a big size spot to perform a low-resolution scan of the entire sensor surface (just a few seconds) firstly. After determining the region of interest (ROI), a small-sized spot is used to scan the local region with a high-resolution. This method is expected to combine the advantages of fast imaging and high-resolution scanning.

In addition, some commercial products suggest illuminating different spots on the LAPS sensor plate, such as a projector based on DLP [42] or liquid crystal [127], an OLED display [125,128]. All the devices are capable of addressing a large number of pixels without mechanical movements, and can customize the measurement areas freely. However, the OLED displays and projectors are designed for video applications, which require low refresh rates (typically 25 fps), so the modulation frequencies in these set-ups are lower than 1 kHz, resulting in a relatively long sampling time, limiting the future applications of fast imaging. Werner et al. developed a new driving method for OLED-LAPS [129], enabled to define the modulation frequencies between 1 kHz and 16 kHz. The increased frequency modulation light speeds up the measurement and reduces the acquisition time of image by a factor of 40 compared with ordinary OLED display [128].

4.2. Multi-Frequency Modulation Light Source Array

Another method for improving the temporal resolution is the utilization of light source array, illuminating multi pixels simultaneously with different modulated frequencies. The fast Fourier transform (FFT) algorithm is used to separate and reconstruct the signal for each frequency from the mixed signal. In order to successfully separate each signal in the frequency domain, it is worth noting that the step of modulation frequency must be larger than or equal to the inverse of the sampling time [21,130,131], and the highest frequency is required to be less than twice the lowest frequency to avoid the harmonic interference [19,131]. Zhang et al. proposed this method for the first time in 2001 [132], the LAPS sensor plate was illuminated simultaneously by 3 light spots at different pixels with the frequencies of 3.6, 3.8 and 4.0 kHz. Later, this method similar to the frequency-division multiplexing (FDM) has been applied for rapid imaging system and multi-sensor with multiple light sources. A light source array consists of 4 × 4 LED for multi-sensor was employed [68], and demonstrated the possibility of recording cell metabolism and ion detection simultaneously at multiple spots on one chip. Hu et al. [69] reported three parallel measurements using line-array focused laser for heavy metal detection. Miyamoto et al. [130] proposed a combination of light source array and the traditional mechanical scanning. A linear array of 16 LEDs with an interval of 3.6 mm was used at different frequencies, the scan speed of this linear LED array was 12.5 mm/s and obtain a image at a resolution of 16 pixels × 128 lines within 6.4 s. In this method, increasing the density of the LED array allows for more measurement spots, and higher resolution can be obtained at the same sampling time. Wagner et al. [133] applied a high-density multi-point LAPS based on a miniaturized vertical-cavity surface-emitting laser (VCSEL) array. It consisted of 12 VCSEL diodes with a pitch of 250 μm on a 3 mm single substrate, and the modulation frequencies of each diode were generated by the FPGA in the range of 3 kHz to 4.1 kHz, with a step of 100 Hz. This set-up could operate measurement spots simultaneously and reduce the overall measurement time.

For LAPS imaging applications, the high modulation frequency requires a lower sampling time, which contributes to the improvement of image speed. In the backside illumination LAPS, the highest modulation frequency is limited by the low-pass filtering characteristics of photocarriers diffusion across the semiconductor substrate [97]. The available modulation frequency can be therefore extended by using frontside illumination or a thinner semiconductor substrate, which can achieve a higher imaging rate. A front-side-illuminated LAPS with two-dimensional LED array was proposed [134]. With the sampling frequency of 400 kHz, the visualization of pH distribution could be recorded at the rate of 70 fps. After that, a high-speed chemical imaging system [131] based on 64 illumination spots guided by optical fibers was developed for dynamic pH measurement inside a microfluidic channel on the LAPS chip surface. Figure 7a shows the schematic of the LAPS with microfluidic channel, the light source array is above the LAPS and performs a liner scan. The photo of 64 light beams guided with optical fibers are shown in Figure 7b, the modulation frequencies are in the range of 6.4 kHz to 12.7 kHz with the step of 100 Hz. In the application of pH distribution visualization, the optical fibers were bundled in a layout of 8 × 8 with an interval of 1.5 mm, and an imaging area of 12 × 12 mm^2 was acquired. The sampling frequency, the sampling number, and the sampling time were 400 kHz, 4000, and 10 ms respectively, which achieved a high-speed imaging at 100 fps. 10 mM NaOH and 10 mM HCl was respectively injected into the phosphate buffered saline (PBS), the frames of the spatiotemporal change in pH distribution are shown in Figure 7c,d, the changes in photocurrent correspond to the changes in pH. A temporary change in pH and subsequent recovery can be clearly observed. All the technical information about temporal resolution mentioned above is summarized in Table 3.

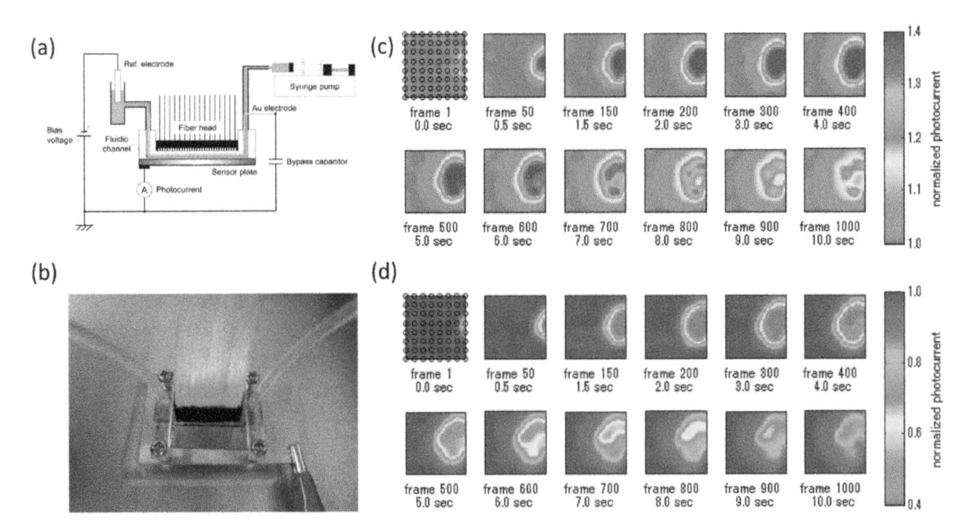

Figure 7. (**a**) Schematic of the LAPS with microfluidic channel set-up; (**b**) The 64 light beams guided by the optical fibers above the LAPS surface; (**c**) Injection of 10 mM NaOH and (**d**) 10 mM HCl into PBS solution. Reprinted (adapted) with permission from Miyamoto et al. [131] Copyright (2019) Elsevier.

Table 3. Technical information about improving the temporal resolution of LAPS.

Methods		Notes	Sensor Construction	Modulated Light Parameters [1]	Temporal Resolution [2]	Reference
Single Modulated Light Without Mechanical Movement	Analog Gimbal-Less Two-Axis Micromirror	Light Spot Movement by Angular Rotation	10 nm Si_3N_4/3 nm SiO_2/500 μm Si	Backside Illumination; λ = 658 nm, f = 5~20 kHz, S = 300 μm	R = 500 × 400 pixels, S1 = 14.5 × 10.5 mm^2, S2 = 300 μm, t = 40 s; R = 10 × 8 pixels, S1 = 2.8 × 5 mm^2, S2 = 300 μm, FPS = 16	[123,124]
	DLP-Based Digital Micromirror Device (DMD)	480 × 320 Micromirror Array; Modulation by Digital Switch	Si_3N_4/SiO_2/Si/Au	Backside Illumination; f = 713 Hz, S = 4.3 μm	S1 = 20.8 × 15.6 mm^2, S2 = 2.6 × 2.6 mm^2, t = 2 s; S1 = 5 × 5 mm^2, S2 = 0.87 × 0.87 mm^2, t = 5 s; S1 = 1 × 1 mm^2, S2 = 0.13 × 0.13 mm^2, t = 60 s	[125,126]
	DLP-Based Projector	\	20 nm HfO_2/1 μm a-Si/10 nm Mo/70 nm ITO/Glass	Backside Illumination; f = 30 Hz, S = 72 μm × 72 μm	R = 160 ×25 pixels, S1 = 2.88 × 1.8 mm^2, S2 = 155× 155 $μm^2$, t = 800 s; R = 98 × 22 pixels, S1 = 1.764 × 1.188 mm^2, S2 = 106× 106 $μm^2$, t = 431.2 s;	[42]
	OLED Display	High Modulation Frequency	Si_3N_4/SiO_2/Si/Al	Backside Illumination; f = 1.74 kHz, S = 200 × 200 μm	R = 96 × 64 pixels, S1 = 20.1 × 13.2 mm^2, S2 = 0.4 × 0.2 mm^2, t = 150 s	[129]
Multi-Frequency Modulation Light Source Array (FDM)	High-Density VCSEL Array	12 VCSEL Diodes with a Pitch of 250 μm	Ta_2O_5/SiO_2/Si	Backside Illumination; λ = 850 nm, f = 3 kHz ~ 4.1 kHz, Step = 100 Hz, S = 500 μm	R = 12 × 22 pixels, S1 = 3 × 10 mm^2, S2 = 0.5 mm × 3 mm, t = 3.6 s	[133]
	Two-Dimensional LED Array	7 × 5 LED Array; Illumination Line by Line	100 nm Si_3N_4/50 nm SiO_2/200 μm Si	Frontside Illumination; λ = 660 nm, f = 6 ~ 10 kHz, Step = 1 kHz, S = 2 mm	R = 7 × 5 pixels, S1 = 17 × 12 mm^2, S2 = 2 mm × 12 mm, FPS = 70	[134]
	Optical Fiber Array with Microfluidic Channel	64 Light Beams; Flexible Measurement Layout	PDMS/Si_3N_4/SiO_2/Si	Frontside Illumination; λ = 600 ~ 625 nm, f = 6.4 ~ 12.7 kHz, Step = 100 Hz, S = 500 μm	R = 8 × 8 pixels, S1 = 12 × 12 mm^2, S2 = 0.5 mm × 0.5 mm, FPS = 100	[131]

[1] λ: Wavelength; f: Frequency; S: Light spot size. [2] R: Image resolution; S1: Image area; S2: Measurement spot size; t: Imaging time; FPS: Frames per second.

5. Integration with Microfluidic Devices

For some biomedical applications with small sample volumes, integrating the sensors with microfluidic devices is a promising strategy. The LAPS-based imaging sensor is a well-suited sensor for the combination with microfluidic devices [37,131,135] as the following advantages:

Firstly, the LAPS chip surface is flat and uniform, which facilitates its fit with microfluidic channels [136] and microchambers [137] of any shape. The commonly used materials for the sensor chip (silicon oxynitride) and microfluidic channel devices (polydimethylsiloxane (PDMS)) can be easily bonded together by plasma treatment. Secondly, due to the addressability of LAPS, any position within the microfluidic channel is measurable [138], which contributes to the assays based on chemical imaging. Thirdly, thanks to the good breathability, biocompatibility and light transmission of PDMS materials, LAPS integrated with microfluidics can be used for biomedical measurements [139] and optical images can also be obtained under a microscope.

LAPS-based chemical imaging sensor can simultaneously analyze the chemical reaction and diffusion of the analyte in a microfluidic channel. Miyamoto et al. [138] proposed a plug-based microfluidic system combined with LAPS. The structure of the integrated system is shown in Figure 8a. The analyte solution was injected into the channel as a plug form with the volume of microliters. Because of the light-addressability, the plug could be monitored at any position and pneumatically controlled. Figure 8b shows the microfluidic channel design to generate plugs. The analyte solution in the sample chamber was divided into a series of plugs by the injected air and then passed through the sensing area.

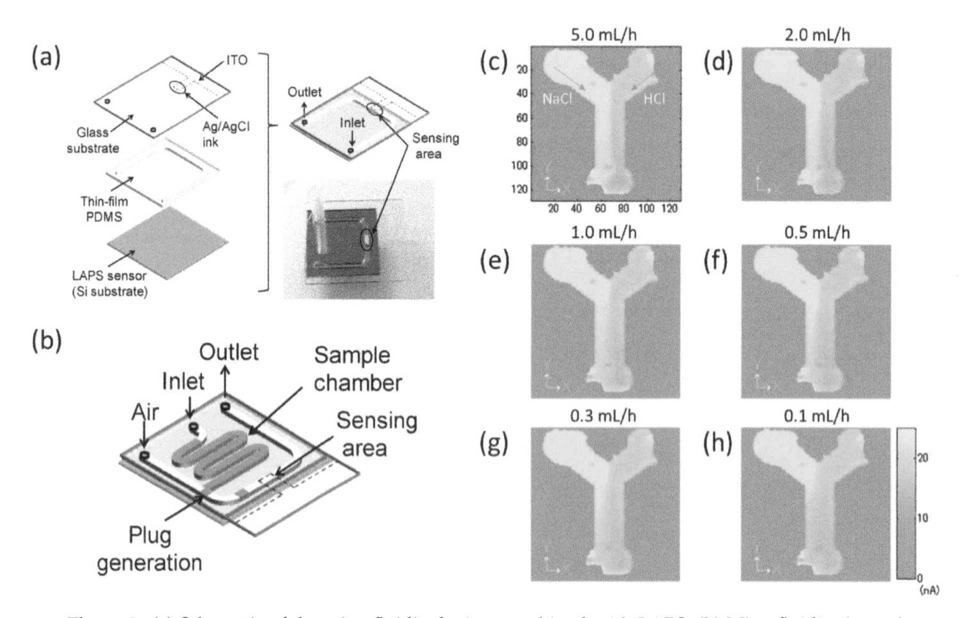

Figure 8. (**a**) Schematic of the microfluidic devices combined with LAPS; (**b**) Microfluidic channel design for plugs generation on LAPS chip. Reprinted from; Chemical images of the laminar flows in the Y-shaped channel under the injection rates from (**c**) 5.0 mL/h to (**h**) 0.1 mL/h. Reprinted (adapted) with permission from Miyamoto et al. [138] and Miyamoto et al. [37] Copyright (2019) Elsevier.

The LAPS-based chemical imaging sensor has also been applied to investigate the concentration profile of ion diffusion in a microfluidic channel [37]. A Y-shaped microfluidic channel was fabricated on the LAPS chip surface, with the thickness and the width of 160 μm and 2 mm. 0.1 M NaCl and 0.1 M HCl were injected respectively into the left and the right branches of the Y-shaped channel, and the chemical image of laminar flows were obtained under the injection rates from 5 mL/h to 0.1 mL/h, as shown in Figure 8c–h. A boundary of the two streams can be clearly seen in Figure 8c because the contacting time is too short and there is no significant change of the pH distribution. As the flow rate decreasing, more ions diffusion across the interface of the two laminar flows can be observed, from which the diffusion coefficient can be calculated. Since the diffusion coefficient depends on the molecular weight of chemical species, this assay can be used as a mass spectrometric measurement of an unknown molecule.

6. Conclusions

Since LAPS was first proposed in 1988, it has been widely developed for biomedical and chemical studies. LAPS are suitable for detecting potential changes induced by changes in ion concentrations, and can also be applied for the measurement of local impedance (SPIM). The flat and uniform surface of LAPS chip facilitates the deposition of sensitive materials and the immobilization of specific identification elements, detections for metal and inorganic salt ions, biomolecules such as enzyme, antibody, glucose and DNA, etc. have been achieved so far.

The performance of LAPS chemical imaging has also enhanced over the past two decades. For high spatial resolution imaging, the LAPS substrate materials including thinned silicon, amorphous silicon (a-Si), and ultra-thin semiconductor layers on transparent substrates have been investigated. The effect of the modulated light property on spatial resolution has also been studied. The best spatial resolution reported to date is 0.8 μm obtained with a SOS device based on the two-photon effect. Based on the photoelectrochemical imaging system (PEIS) with ITO-LAPS, the visualization extent of the metabolism and lysis can be accurate to the single cell level, which expands the application prospects of LAPS imaging.

The use of a FDM-based multi-light source array is an effective method for high-speed imaging. Parallel measurement of multiple pixels allows signals of tens of thousands of pixels to be read out in one second. This high temporal resolution is able to measure the processes of fast chemical reactions such as ion diffusion. A variety of measurement modes were designed to meet different needs, and image calibration algorithms were developed to correct the non-uniformities of sensor chip.

For further application of LAPS, in addition to improving the sensor performance, the miniaturization multi-functionalization of the detection system is also important. It is a promising prospect that setting multiple measurement sites on the sensor surface and integrating with the microfluidic device to establish a "lab-on-a-chip". On one hand, the volume of sample in microliter grade can reduce costs; on the other hand, it enables rapid and multi-parameter detection under controllable dynamic conditions with the emergence of novel designed microfluidic devices [140]. The integrated sensor system will be able to be applied in biomedical applications for a variety of purposes.

Author Contributions: Conceptualization, investigation and writing - original draft preparation, T.L.; resources, Y.Q. and Y.G.; supervision, J.S.; visualization, S.Z.; project administration, H.W.; funding acquisition, P.W.

Funding: This research was funded by Major State Basic Research Development Program of China (973 Program), grant number 2015CB352101; National Natural Science Foundation of China, grant number 31571004, 31661143030, 51861145307; International Cooperation and Exchange Program (NSFC-RFBR), grant number 8171101322; Fundamental Research Funds for the Central Universities, grant number 2018QNA5018, 2018FZA5018; Major Research and Development Project of Zhejiang Province, grant number 2017C03032, 2019C03066.

Acknowledgments: Thanks to Prof. Chunsheng Wu and Liping Du from Xi'an Jiaotong University, and the colleagues Chenlei Gu and Xinyi Wang from Zhejiang University for their help in completing this review.

Conflicts of Interest: The authors declare no conflict of interest. The funders had no role in the design of the study; in the collection, analyses, or interpretation of data; in the writing of the manuscript, or in the decision to publish the results.

References

1. Hafeman, D.G.; Parce, J.W.; McConnell, H.M. Light-addressable potentiometric sensor for biochemical systems. *Science* **1988**, *240*, 1182–5118. [CrossRef] [PubMed]
2. McConnell, H.M.; Owicki, J.C.; Parce, J.W.; Miller, D.L.; Baxter, G.T.; Wada, H.G.; Pitchford, S. The cytosensor microphysiometer: Biological applications of silicon technology. *Science* **1992**, *257*, 1906–7912. [CrossRef] [PubMed]

3. Owicki, J.C.; Bousse, L.J.; Hafeman, D.G.; Kirk, G.L.; Olson, J.D.; Wada, H.G.; Parce, J.W. The light-addressable potentiometric sensor: Principles and biological applications. *Annu. Rev. Biophys. Biomol. Struct.* **1994**, *23*, 87–113. [CrossRef] [PubMed]
4. Hafner, F. Cytosensor Microphysiometer: Technology and recent applications. *Biosens. Bioelectron.* **2000**, *15*, 149–158. [CrossRef]
5. Stein, B.; George, M.; Gaub, H.E.; Behrends, J.C.; Parak, W.J. Spatially resolved monitoring of cellular metabolic activity with a semiconductor-based biosensor. *Biosens. Bioelectron.* **2003**, *18*, 31–41. [CrossRef]
6. Wagner, T.; Schöning, M.J. Light-addressable potentiometric sensors (LAPS): Recent trends and applications. *Compr. Anal. Chem.* **2007**, *49*, 87–128.
7. Bratov, A.; Abramova, N.; Ipatov, A. Recent trends in potentiometric sensor arrays—A review. *Anal. Chim. Acta* **2010**, *678*, 149–159. [CrossRef]
8. Schöning, M.J.; Poghossian, A. Bio FEDs (Field-Effect devices): State-of-the-art and new directions. *Electroanalysis* **2006**, *18*, 1893–1900. [CrossRef]
9. Vlasov, Y.G.; Tarantov, Y.A.; Bobrov, P.V. Analytical characteristics and sensitivity mechanisms of electrolyte-insulator-semiconductor system-based chemical sensors-a critical review. *Anal. Bioanal. Chem.* **2003**, *376*, 788–796. [CrossRef]
10. Schöning, M.J.; Poghossian, A.; Yoshinobu, T.; Luth, H. Semiconductor-based field-effect structures for chemical sensing. *Adv. Environ. Chem. Sens. Technol.* **2001**, *4205*, 188–198.
11. Poghossian, A.; Yoshinobu, T.; Simonis, A.; Ecken, H.; Luth, H.; Schöning, M.J. Penicillin detection by means of field-effect based sensors: EnFET, capacitive EIS sensor or LAPS? *Sens. Actuators B Chem.* **2001**, *78*, 237–242. [CrossRef]
12. Verzellesi, G.; Colalongo, L.; Passeri, D.; Margesin, B.; Rudan, M.; Soncini, G.; Ciampolini, P. Numerical analysis of ISFET and LAPS devices. *Sens. Actuators B Chem.* **1997**, *44*, 402–408. [CrossRef]
13. Poghossian, A.; Ingebrandt, S.; Offenhausser, A.; Schöning, M.J. Field-effect devices for detecting cellular signals. *Semin. Cell. Dev. Biol.* **2009**, *20*, 41–48. [CrossRef] [PubMed]
14. Choi, K.; Kim, J.Y.; Ahn, J.H.; Choi, J.M.; Im, M.; Choi, Y.K. Integration of field effect transistor-based biosensors with a digital microfluidic device for a lab-on-a-chip application. *Lab Chip* **2012**, *12*, 1533–1539. [CrossRef] [PubMed]
15. Poghossian, A.; Schöning, M.J. Label-free sensing of biomolecules with field-effect devices for clinical applications. *Electroanalysis* **2014**, *26*, 1197–1213. [CrossRef]
16. Zhou, Y.; Chen, L.; Krause, S.; Chazalviel, J.N. Scanning photoinduced impedance microscopy using amorphous silicon photodiode structures. *Anal. Chem.* **2007**, *79*, 6208–6214. [CrossRef]
17. Krause, S.; Moritz, W.; Talabani, H.; Xu, M.; Sabot, A.; Ensell, G. Scanning photo-induced impedance microscopy-resolution studies and polymer characterization. *Electrochim. Acta* **2006**, *51*, 1423–1430. [CrossRef]
18. Krause, S.; Talabani, H.; Xu, M.; Moritz, W.; Griffiths, J. Scanning photo-induced impedance microscopy–An impedance based imaging technique. *Electrochim. Acta* **2002**, *47*, 2143–2148. [CrossRef]
19. Yoshinobu, T.; Miyamoto, K.; Wagner, T.; Schöning, M.J. Recent developments of chemical imaging sensor systems based on the principle of the light-addressable potentiometric sensor. *Sens. Actuators B Chem.* **2015**, *207*, 926–932. [CrossRef]
20. Wu, F.; Campos, I.; Zhang, D.W.; Krause, S. Biological imaging using light-addressable potentiometric sensors and scanning photo-induced impedance microscopy. *Proc. Math. Phys. Eng. Sci.* **2017**, *473*, 20170130. [CrossRef]
21. Yoshinobu, T.; Miyamoto, K.; Werner, C.F.; Poghossian, A.; Wagner, T.; Schöning, M.J. Light-addressable potentiometric sensors for quantitative spatial imaging of chemical species. *Annu. Rev. Anal. Chem.* **2017**, *10*, 225–246. [CrossRef] [PubMed]
22. Wang, J.; Du, L.P.; Krause, S.; Wu, C.S.; Wang, P. Surface modification and construction of LAPS towards biosensing applications. *Sens. Actuators B Chem.* **2018**, *265*, 161–173. [CrossRef]
23. Liang, J.T.; Guan, M.Y.; Huang, G.Y.; Qiu, H.M.; Chen, Z.C.; Li, G.Y.; Huang, Y. Highly sensitive covalently functionalized light-addressable potentiometric sensor for determination of biomarker. *Mat. Sci. Eng. C Mater.* **2016**, *63*, 185–191. [CrossRef] [PubMed]
24. Liang, J.T.; Zhu, N.X.; Li, S.S.; Jia, H.Q.; Xue, Y.W.; Cui, L.J.; Huang, Y.; Li, G.Y. Light-addressable potentiometric sensor with gold nanoparticles enhancing enzymatic silver deposition for 1,5-anhydroglucitol determination. *Biochem. Eng. J.* **2017**, *123*, 29–37. [CrossRef]

25. Miyamoto, K.; Yu, B.; Isoda, H.; Wagner, T.; Schöning, M.J.; Yoshinobu, T. Visualization of the recovery process of defects in a cultured cell layer by chemical imaging sensor. *Sens. Actuators B Chem.* **2016**, *236*, 965–969. [CrossRef]
26. Zhang, D.W.; Wu, F.; Wang, J.; Watkinson, M.; Krause, S. Image detection of yeast *Saccharomyces cerevisiae* by light-addressable potentiometric sensors (LAPS). *Electrochem. Commun.* **2016**, *72*, 41–45. [CrossRef]
27. Miyamoto, K.; Sakakita, S.; Yoshinobu, T. A novel data acquisition method for visualization of large pH changes by chemical imaging sensor. *ISIJ Int.* **2016**, *56*, 492–494. [CrossRef]
28. Nakao, M.; Yoshinobu, T.; Iwasaki, H. Scanning-laser-beam semiconductor pH-imaging sensor. *Sens. Actuators B Chem.* **1994**, *20*, 119–123. [CrossRef]
29. Nakao, M.; Yoshinobu, T.; Iwasaki, H. Improvement of spatial-resolution of a laser-scanning pH-imaging sensor. *Jpn. J. Appl. Phys.* **1994**, *33*, L394–L397. [CrossRef]
30. Nakao, M.; Inoue, S.; Oishi, R.; Yoshinobu, T.; Iwasaki, H. Observation of microorganism colonies using a scanning-laser-beam pH-sensing microscope. *J. Ferment. Bioeng.* **1995**, *79*, 163–166. [CrossRef]
31. Nakao, M.; Inoue, S.; Yoshinobu, T.; Iwasaki, H. High-resolution pH imaging sensor for microscopic observation of microorganisms. *Sens. Actuators B Chem.* **1996**, *34*, 234–239. [CrossRef]
32. Dantism, S.; Takenaga, S.; Wagner, T.; Wagner, P.; Schöning, M.J. Differential imaging of the metabolism of bacteria and eukaryotic cells based on light-addressable potentiometric sensors. *Electrochim. Acta* **2017**, *246*, 234–241. [CrossRef]
33. Miyamoto, K.I.; Sugawara, Y.; Kanoh, S.; Yoshinobu, T.; Wagner, T.; Schöning, M.J. Image correction method for the chemical imaging sensor. *Sens. Actuators B Chem.* **2010**, *144*, 344–348. [CrossRef]
34. Inoue, S.; Nakao, M.; Yoshinobu, T.; Iwasaki, H. Chemical-imaging sensor using enzyme. *Sens. Actuators B Chem.* **1996**, *32*, 23–26. [CrossRef]
35. Wang, J.; Campos, I.; Wu, F.; Zhu, J.Y.; Sukhorukov, G.B.; Palma, M.; Watkinson, M.; Krause, S. The effect of gold nanoparticles on the impedance of microcapsules visualized by scanning photo-induced impedance microscopy. *Electrochim. Acta* **2016**, *208*, 39–46. [CrossRef]
36. Yoshinobu, T.; Harada, T.; Iwasaki, H. Application of the pH-imaging sensor to determining the diffusion coefficients of ions in electrolytic solutions. *Jpn. J. Appl. Phys.* **2000**, *39*, L318–L320. [CrossRef]
37. Miyamoto, K.; Ichimura, H.; Wagner, T.; Schöning, M.J.; Yoshinobu, T. Chemical imaging of the concentration profile of ion diffusion in a microfluidic channel. *Sens. Actuators B Chem.* **2013**, *189*, 240–245. [CrossRef]
38. Hu, N.; Wu, C.; Ha, D.; Wang, T.; Liu, Q.; Wang, P. A novel microphysiometer based on high sensitivity LAPS and microfluidic system for cellular metabolism study and rapid drug screening. *Biosens. Bioelectron.* **2013**, *40*, 167–173. [CrossRef] [PubMed]
39. Wan, H.; Ha, D.; Zhang, W.; Zhao, H.X.; Wang, X.; Sun, Q.Y.; Wang, P. Design of a novel hybrid sensor with microelectrode array and LAPS for heavy metal determination using multivariate nonlinear calibration. *Sens. Actuators B Chem.* **2014**, *192*, 755–761. [CrossRef]
40. Yang, C.M.; Zeng, W.Y.; Chen, C.H.; Chen, Y.P.; Chen, T.C. Spatial resolution and 2D chemical image of light-addressable potentiometric sensor improved by inductively coupled-plasma reactive-ion etching. *Sens. Actuators B Chem.* **2018**, *258*, 1295–1301. [CrossRef]
41. Moritz, W.; Yoshinobu, T.; Finger, F.; Krause, S.; Martin-Fernandez, M.; Schöning, M.J. High resolution LAPS using amorphous silicon as the semiconductor material. *Sens. Actuators B Chem.* **2004**, *103*, 436–441. [CrossRef]
42. Das, A.; Lin, Y.H.; Lai, C.S. Miniaturized amorphous-silicon based chemical imaging sensor system using a mini-projector as a simplified light-addressable scanning source. *Sens. Actuators B Chem.* **2014**, *190*, 664–672. [CrossRef]
43. Yang, C.M.; Liao, Y.H.; Chen, C.H.; Chen, T.C.; Lai, C.S.; Pijanowska, D.G. P-I-N amorphous silicon for thin-film light-addressable potentiometric sensors. *Sens. Actuators B Chem.* **2016**, *236*, 1005–1010. [CrossRef]
44. Yoshinobu, T.; Schöning, M.J.; Finger, F.; Moritz, W.; Iwasaki, H. Fabrication of thin-film LAPS with amorphous silicon. *Sensors* **2004**, *4*, 163–169. [CrossRef]
45. Chen, L.; Zhou, Y.L.; Jiang, S.H.; Kunze, J.; Schmuki, P.; Krause, S. High resolution LAPS and SPIM. *Electrochem. Commun.* **2010**, *12*, 758–760. [CrossRef]
46. Chen, L.; Zhou, Y.L.; Krause, S.; Munoz, A.G.; Kunze, J.; Schmuki, P. Repair of thin thermally grown silicon dioxide by anodic oxidation. *Electrochim. Acta* **2008**, *53*, 3395–3402. [CrossRef]

47. Moritz, W.; Gerhardt, I.; Roden, D.; Xu, M.; Krause, S. Photocurrent measurements for laterally resolved interface characterization. *Fresenius J. Anal. Chem.* **2000**, *367*, 329–333. [CrossRef]
48. Das, A.; Chang, L.B.; Lai, C.S.; Lin, R.M.; Chu, F.C.; Lin, Y.H.; Chow, L.; Jeng, M.J. GaN thin film based light addressable potentiometric sensor for pH sensing application. *Appl. Phys. Express* **2013**, *6*, 036601. [CrossRef]
49. Yang, C.M.; Chen, C.H.; Chang, L.B.; Lai, C.S. IGZO thin-film light-addressable potentiometric sensor. *IEEE Electr. Device L* **2016**, *37*, 1481–1484. [CrossRef]
50. Zhang, D.W.; Wu, F.; Krause, S. LAPS and SPIM imaging using ITO-coated glass as the substrate material. *Anal. Chem.* **2017**, *89*, 8129–8133. [CrossRef]
51. Yoshinobu, T.; Ecken, H.; Poghossian, A.; Luth, H.; Iwasaki, H.; Schöning, M.J. Alternative sensor materials for light-addressable potentiometric sensors. *Sens. Actuators B Chem.* **2001**, *76*, 388–392. [CrossRef]
52. Schöning, M.J.; Tsarouchas, D.; Beckers, L.; Schubert, J.; Zander, W.; Kordos, P.; Luth, H. A highly long-term stable silicon-based pH sensor fabricated by pulsed laser deposition technique. *Sens. Actuators B Chem.* **1996**, *35*, 228–233. [CrossRef]
53. Seki, A.; Motoya, K.; Watanabe, S.; Kubo, I. Novel sensors for potassium, calcium and magnesium ions based on a silicon transducer as a light-addressable potentiometric sensor. *Anal. Chim. Acta* **1999**, *382*, 131–136. [CrossRef]
54. Ismail, A.M.; Harada, T.; Yoshinobu, T.; Iwasaki, H.; Schöning, M.J.; Luth, H. Investigation of pulsed laser-deposited Al_2O_3 as a high pH-sensitive layer for LAPS-based biosensing applications. *Sens. Actuators B Chem.* **2000**, *71*, 169–172. [CrossRef]
55. Lue, C.E.; Lai, C.S.; Chen, H.Y.; Yang, C.M. Light addressable potentiometric sensor with fluorine-terminated hafnium oxide layer for sodium detection. *Jpn. J. Appl. Phys.* **2010**, *49*, 04DL05. [CrossRef]
56. Chin, C.H.; Lu, T.F.; Wang, J.C.; Yang, J.H.; Lue, C.E.; Yang, C.M.; Li, S.S.; Lai, C.S. Effects of CF4 plasma treatment on pH and pNa sensing properties of light-addressable potentiometric sensor with a 2-nm-thick sensitive HfO_2 layer grown by atomic layer deposition. *Jpn. J. Appl. Phys.* **2011**, *50*, 04DL06. [CrossRef]
57. Yang, J.H.; Lu, T.F.; Wang, J.C.; Yang, C.M.; Pijanowska, D.G.; Chin, C.H.; Lue, C.E.; Lai, C.S. LAPS with nanoscaled and highly polarized HfO_2 by CF_4 plasma for $NH_4(+)$ detection. *Sens. Actuators B Chem.* **2013**, *180*, 71–76. [CrossRef]
58. Chen, C.H.; Yang, C.M. A IGZO-based light-addressable potentiometric sensor on a PET susbtrate. In Proceedings of the 2019 IEEE International Conference on Flexible and Printable Sensors and Systems (FLEPS), Glasgow, UK, 8–10 July 2019; pp. 1–2.
59. Yang, C.M.; Chiang, T.W.; Yeh, Y.T.; Das, A.; Lin, Y.T.; Chen, T.C. Sensing and pH-imaging properties of niobium oxide prepared by rapid thermal annealing for electrolyte-insulator-semiconductor structure and light-addressable potentiometric sensor. *Sens. Actuators B Chem.* **2015**, *207*, 858–864. [CrossRef]
60. Yang, C.M.; Yang, Y.C.; Chen, C.H. Thin-film light-addressable potentiometric sensor with SnOx as a photosensitive semiconductor. *Vacuum* **2019**, *168*, 108809. [CrossRef]
61. Wang, J.; Zhou, Y.L.; Watkinson, M.; Gautrot, J.; Krause, S. High-sensitivity light-addressable potentiometric sensors using silicon on sapphire functionalized with self-assembled organic monolayers. *Sens. Actuators B Chem.* **2015**, *209*, 230–236. [CrossRef]
62. Yang, Y.; Cuartero, M.; Goncales, V.R.; Gooding, J.J.; Bakker, E. Light-addressable ion sensing for real-time monitoring of extracellular potassium. *Angew. Chem. Int. Ed. Engl.* **2018**, *57*, 16801–16805. [CrossRef] [PubMed]
63. Shaibani, P.M.; Jiang, K.R.; Haghighat, G.; Hassanpourfard, M.; Etayash, H.; Naicker, S.; Thundat, T. The detection of *Escherichia coli* (*E. coli*) with the pH sensitive hydrogel nanofiber-light addressable potentiometric sensor (NF-LAPS). *Sens. Actuators B Chem.* **2016**, *226*, 176–183. [CrossRef]
64. Shaibani, P.M.; Etayash, H.; Naicker, S.; Kaur, K.; Thundat, T. Metabolic study of cancer cells using a pH sensitive hydrogel nanofiber light addressable potentiometric sensor. *ACS Sens.* **2017**, *2*, 151–156. [CrossRef] [PubMed]
65. Mourzina, Y.; Yoshinobu, T.; Schubert, J.; Luth, H.; Iwasaki, H.; Schöning, M.J. Ion-selective light-addressable potentiometric sensor (LAPS) with chalcogenide thin film prepared by pulsed laser deposition. *Sens. Actuators B Chem.* **2001**, *80*, 136–140. [CrossRef]
66. Men, H.; Zou, S.F.; Li, Y.; Wang, Y.P.; Ye, X.S.; Wang, P. A novel electronic tongue combined MLAPS with stripping voltammetry for environmental detection. *Sens. Actuators B Chem.* **2005**, *110*, 350–357. [CrossRef]

67. Kloock, J.P.; Moreno, L.; Bratov, A.; Huachupoma, S.; Xu, J.; Wagner, T.; Yoshinobu, T.; Ermolenko, Y.; Vlasov, Y.G.; Schöning, M.J. PLD-prepared cadmium sensors based on chalcogenide glasses–ISFET, LAPS and mu ISE semiconductor structures. *Sens. Actuators B Chem.* **2006**, *118*, 149–155. [CrossRef]
68. Wagner, T.; Molina, R.; Yoshinobu, T.; Kloock, J.P.; Biselli, M.E.; Canzoneri, M.; Schnitzler, T.; Schöning, M.J. Handheld multi-channel LAPS device as a transducer platform for possible biological and chemical multi-sensor applications. *Electrochim. Acta* **2007**, *53*, 305–311. [CrossRef]
69. Hu, W.; Cai, H.; Fu, J.; Wang, P.; Yang, G. Line-scanning LAPS array for measurement of heavy metal ions with micro-lens array based on MEMS. *Sens. Actuators B Chem.* **2008**, *129*, 397–403. [CrossRef]
70. Ha, D.; Hu, N.; Wu, C.X.; Kirsanov, D.; Legin, A.; Khaydukova, M.; Wang, P. Novel structured light-addressable potentiometric sensor array based on PVC membrane for determination of heavy metals. *Sens. Actuators B Chem.* **2012**, *174*, 59–64. [CrossRef]
71. Ermolenko, Y.; Yoshinobu, T.; Mourzina, Y.; Furuichi, K.; Levichev, S.; Vlasov, Y.; Schöning, M.J.; Iwasaki, H. Lithium sensor based on the laser scanning semiconductor transducer. *Anal. Chim. Acta* **2002**, *459*, 1–9. [CrossRef]
72. Yoshinobu, T.; Schöning, M.J.; Otto, R.; Furuichi, K.; Mourizina, Y.; Ermolenko, Y.; Iwasaki, H. Portable light-addressable potentiometric sensor (LAPS) for multisensor applications. *Sens. Actuators B Chem.* **2003**, *95*, 352–356. [CrossRef]
73. Mourzina, Y.G.; Ermolenko, Y.E.; Yoshinobu, T.; Vlasov, Y.; Iwasaki, H.; Schöning, M.J. Anion-selective light-addressable potentiometric sensors (LAPS) for the determination of nitrate and sulphate ions. *Sens. Actuators B Chem.* **2003**, *91*, 32–38. [CrossRef]
74. Wu, Y.C.; Wang, P.; Ye, X.S.; Zhang, Q.T.; Li, R.; Yan, W.M.; Zheng, X.X. A novel microphysiometer based on MLAPS for drugs screening. *Biosens. Bioelectron.* **2001**, *16*, 277–286.
75. Yoshinobu, T.; Iwasaki, H.; Ui, Y.; Furuichi, K.; Ermolenko, Y.; Mourzina, Y.; Wagner, T.; Nather, N.; Schöning, M.J. The light-addressable potentiometric sensor for multi-ion sensing and imaging. *Methods* **2005**, *37*, 94–102. [CrossRef] [PubMed]
76. Ermolenko, Y.E.; Yoshinobu, T.; Mourzina, Y.G.; Vlasov, Y.G.; Schöning, M.J.; Iwasaki, H. Laser-scanned silicon transducer (LSST) as a multisensor system. *Sens. Actuators B Chem.* **2004**, *103*, 457–462. [CrossRef]
77. Tsujimura, Y.; Yokoyama, M.; Kimura, K. Comparison between silicone-rubber membranes and plasticized poly (vinyl chloride) membranes containing calix [4] arene ionophores for sodium ion-sensitive field-effect transistors in applicability to sodium assay in human-body fluids. *Sens. Actuators B Chem.* **1994**, *22*, 195–199. [CrossRef]
78. Kimura, K.; Tsujimura, Y.; Yokoyama, M. Silicone-rubber membrane sodium-ion sensors based on calix [4] arene neutral carriers. *Pure Appl. Chem.* **1995**, *67*, 1085–1089. [CrossRef]
79. Makarychev-Mikhailov, S.; Legin, A.; Mortensen, J.; Levitchev, S.; Vlasov, Y. Potentiometric and theoretical studies of the carbonate sensors based on 3-bromo-4-hexyl-5-nitrotrifluoroacetophenone. *Analyst* **2004**, *129*, 213–218. [CrossRef]
80. Cazale, A.; Sant, W.; Launay, J.; Ginot, F.; Temple-Boyer, P. Study of field effect transistors for the sodium ion detection using fluoropolysiloxane-based sensitive layers. *Sens. Actuators B Chem.* **2013**, *177*, 515–521. [CrossRef]
81. Cazale, A.; Sant, W.; Ginot, F.; Launay, J.C.; Savourey, G.; Revol-Cavalier, F.; Lagarde, J.M.; Heinry, D.; Launay, J.; Temple-Boyer, P. Physiological stress monitoring using sodium ion potentiometric microsensors for sweat analysis. *Sens. Actuators B Chem.* **2016**, *225*, 1–9. [CrossRef]
82. Ermolenko, Y.; Yoshinobu, T.; Mourzina, Y.; Levichev, S.; Furuichi, K.; Vlasov, Y.; Schöning, M.J.; Iwasaki, H. Photocurable membranes for ion-selective light-addressable potentiometric sensor. *Sens. Actuators B Chem.* **2002**, *85*, 79–85. [CrossRef]
83. Shimizu, M.; Kanai, Y.; Uchida, H.; Katsube, T. Integrated biosensor employing a surface photovoltage technique. *Sens. Actuators B Chem.* **1994**, *20*, 187–192. [CrossRef]
84. Seki, A.; Ikeda, S.; Kubo, I.; Karube, I. Biosensors based on light-addressable potentiometric sensors for urea, penicillin and glucose. *Anal. Chim. Acta* **1998**, *373*, 9–13. [CrossRef]
85. Aoki, K.; Uchida, H.; Katsube, T.; Ishimaru, Y.; Iida, T. Integration of bienzymatic disaccharide sensors for simultaneous determination of disaccharides by means of light addressable potentiometric sensor. *Anal. Chim. Acta* **2002**, *471*, 3–12. [CrossRef]

86. Mourzina, I.G.; Yoshinobu, T.; Ermolenko, Y.E.; Vlasov, Y.G.; Schöning, M.J.; Iwasaki, H. Immobilization of urease and cholinesterase on the surface of semiconductor transducer for the development of light-addressable potentiometric sensors. *Microchim. Acta* **2004**, *144*, 41–50.
87. Siqueira, J.R.; Werner, C.F.; Backer, M.; Poghossian, A.; Zucolotto, V.; Oliveira, O.N.; Schöning, M.J. Layer-by-layer assembly of carbon nanotubes incorporated in light-addressable potentiometric sensors. *J. Phys. Chem. C* **2009**, *113*, 14765–14770. [CrossRef]
88. Siqueira, J.R., Jr.; Maki, R.M.; Paulovich, F.V.; Werner, C.F.; Poghossian, A.; de Oliveira, M.C.; Zucolotto, V.; Oliveira, O.N., Jr.; Schöning, M.J. Use of information visualization methods eliminating cross talk in multiple sensing units investigated for a light-addressable potentiometric sensor. *Anal. Chem.* **2010**, *82*, 61–65. [CrossRef] [PubMed]
89. Miyamoto, K.; Yoshida, M.; Sakai, T.; Matsuzaka, A.; Wagner, T.; Kanoh, S.; Yoshinobu, T.; Schöning, M.J. Differential setup of light-addressable potentiometric sensor with an enzyme reactor in a flow channel. *Jpn. J. Appl. Phys.* **2011**, *50*, 04DL08. [CrossRef]
90. Werner, C.F.; Takenaga, S.; Taki, H.; Sawada, K.; Schöning, M.J. Comparison of label-free ACh-imaging sensors based on CCD and LAPS. *Sens. Actuators B Chem.* **2013**, *177*, 745–752. [CrossRef]
91. Jia, Y.; Gao, C.; Feng, D.; Wu, M.; Liu, Y.; Chen, X.; Xing, K.; Feng, X. Bio-initiated light addressable potentiometric sensor for unlabeled biodetection and its MEDICI simulation. *Analyst* **2011**, *136*, 4533–4538. [CrossRef]
92. Jia, Y.F.; Gao, C.Y.; He, J.; Feng, D.F.; Xing, K.L.; Wu, M.; Liu, Y.; Cai, W.S.; Feng, X.Z. Unlabeled multi tumor marker detection system based on bioinitiated light addressable potentiometric sensor. *Analyst* **2012**, *137*, 3806–3813. [CrossRef]
93. Jia, Y.; Yin, X.B.; Zhang, J.; Zhou, S.; Song, M.; Xing, K.L. Graphene oxide modified light addressable potentiometric sensor and its application for ssDNA monitoring. *Analyst* **2012**, *137*, 5866–5873. [CrossRef] [PubMed]
94. Wu, C.S.; Bronder, T.; Poghossian, A.; Werner, C.F.; Backer, M.; Schöning, M.J. Label-free electrical detection of DNA with a multi-spot LAPS: First step towards light-addressable DNA chips. *Phys. Status Solidi A* **2014**, *211*, 1423–1428. [CrossRef]
95. Wu, C.; Bronder, T.; Poghossian, A.; Werner, C.F.; Schöning, M.J. Label-free detection of DNA using a light-addressable potentiometric sensor modified with a positively charged polyelectrolyte layer. *Nanoscale* **2015**, *7*, 6143–6150. [CrossRef] [PubMed]
96. Wu, C.S.; Poghossian, A.; Bronder, T.S.; Schöning, M.J. Sensing of double-stranded DNA molecules by their intrinsic molecular charge using the light-addressable potentiometric sensor. *Sens. Actuators B Chem.* **2016**, *229*, 506–512. [CrossRef]
97. Sartore, M.; Adami, M.; Nicolini, C.; Bousse, L.; Mostarshed, S.; Hafeman, D. Minority-carrier diffusion length effects on light-addressable potentiometric sensor (laps) devices. *Sens. Actuators A Phys.* **1992**, *32*, 431–436. [CrossRef]
98. Werner, C.F.; Wagner, T.; Yoshinobu, T.; Keusgen, M.; Schöning, M.J. Frequency behaviour of light-addressable potentiometric sensors. *Phys. Status Solidi A* **2013**, *210*, 884–891. [CrossRef]
99. Guo, Y.Y.; Miyamoto, K.; Wagner, T.; Schöning, M.J.; Yoshinobu, T. The oretical study and simulation of light-addressable potentiometric sensors. *Phys. Status Solidi A* **2014**, *211*, 1467–1472. [CrossRef]
100. Guo, Y.; Miyamoto, K.; Wagner, T.; Schöning, M.J.; Yoshinobu, T. Device simulation of the light-addressable potentiometric sensor for the investigation of the spatial resolution. *Sens. Actuators B Chem.* **2014**, *204*, 659–665. [CrossRef]
101. George, M.; Parak, W.J.; Gerhardt, I.; Moritz, W.; Kaesen, F.; Geiger, H.; Eisele, I.; Gaub, H.E. Investigation of the spatial resolution of the light-addressable potentiometric sensor. *Sens. Actuators A Phys.* **2000**, *86*, 187–196. [CrossRef]
102. Parak, W.J.; George, M.; Domke, J.; Radmacher, M.; Behrends, J.C.; Denyer, M.C.; Gaub, H.E. Can the light-addressable potentiometric sensor (LAPS) detect extracellular potentials of cardiac myocytes? *IEEE Trans. Biomed. Eng.* **2000**, *47*, 1106–1113. [CrossRef] [PubMed]
103. Yates, D.E.; Levine, S.; Healy, T.W. Site-binding model of electrical double-layer at oxide-water interface. *J. Chem. Soc. Farad. Trans. 1* **1974**, *70*, 1807–1818. [CrossRef]
104. Bousse, L.; Meindl, J.D. Surface potential-pH characteristics in the theory of the oxide-electrolyte interface. *Geochem. Processes Miner. Surf.* **1986**, *323*, 253–264.

105. Fung, C.D.; Cheung, P.W.; Ko, W.H. A generalized theory of an electrolyte-insulator-semiconductor field-effect transistor. *IEEE Trans. Electron. Devices* **1986**, *33*, 8–18. [CrossRef]
106. Van Hal, R.E.G.; Eijkel, J.C.T.; Bergveld, P. A general model to describe the electrostatic potential at electrolyte oxide interfaces. *Adv. Colloid Interfac.* **1996**, *69*, 31–62. [CrossRef]
107. Yoshinobu, T.; Ecken, H.; Poghossian, A.; Simonis, A.; Iwasaki, H.; Luth, H.; Schöning, M.J. Constant-current-mode LAPS (CLAPS) for the detection of penicillin. *Electroanalysis* **2001**, *13*, 733–736. [CrossRef]
108. Kinameri, K.; Munakata, C.; Mayama, K. A scanning photon microscope for non-destructive observations of crystal defect and interface trap distributions in silicon-wafers. *J. Phys. E Sci. Instrum.* **1988**, *21*, 91–97. [CrossRef]
109. Miyamoto, K.; Wagner, T.; Yoshinobu, T.; Kanoh, S.; Schöning, M.J. Phase-mode LAPS and its application to chemical imaging. *Sens. Actuators B Chem.* **2011**, *154*, 28–32. [CrossRef]
110. Werner, C.F.; Miyamoto, K.; Wagner, T.; Schöning, M.J.; Yoshinobu, T. Lateral resolution enhancement of pulse-driven light-addressable potentiometric sensor. *Sens. Actuators B Chem.* **2017**, *248*, 961–965. [CrossRef]
111. Ito, Y. High-spatial resolution LAPS. *Sens. Actuators B Chem.* **1998**, *52*, 107–111. [CrossRef]
112. Parak, W.J.; Hofmann, U.G.; Gaub, H.E.; Owicki, J.C. Lateral resolution of light-addressable potentiometric sensors: An experimental and theoretical investigation. *Sens. Actuators A Phys.* **1997**, *63*, 47–57. [CrossRef]
113. Zhang, Q.T. The oretical analysis and design of submicron-LAPS. *Sens. Actuators B Chem.* **2005**, *105*, 304–311. [CrossRef]
114. Vandenheuvel, J.C.; Vanoort, R.C.; Geerts, M.J. Diffusion length measurements of thin amorphous-silicon layers. *Solid State Commun.* **1989**, *69*, 807–810. [CrossRef]
115. Wang, J.; Kong, S.; Chen, F.M.; Chen, W.; Du, L.P.; Cai, W.; Huang, L.Q.; Wu, C.S.; Zhang, D.W. A bioelectronic taste sensor based on bioengineered *Escherichia coli* cells combined with ITO-constructed electrochemical sensors. *Anal. Chim. Acta* **2019**, *1079*, 73–78. [CrossRef] [PubMed]
116. Wu, F.; Zhou, B.; Wang, J.; Zhong, M.C.; Das, A.; Watkinson, M.; Hing, K.; Zhang, D.W.; Krause, S. Photoelectrochemical imaging system for the mapping of cell surface charges. *Anal. Chem.* **2019**, *91*, 5896–5903. [CrossRef] [PubMed]
117. Guo, Y.Y.; Seki, K.; Miyamoto, K.; Wagner, T.; Schöning, M.J.; Yoshinobu, T. Novel photoexcitation method for light-addressable potentiometric sensor with higher spatial resolution. *Appl. Phys. Express* **2014**, *7*, 067301. [CrossRef]
118. Guo, Y.Y.; Seki, K.; Miyamoto, K.; Wagner, T.; Schöning, M.J.; Yoshinobu, T. Device simulation of the light-addressable potentiometric sensor with a novel photoexcitation method for a higher spatial resolution. In Proceedings of the 28th European Conference on Solid-State Transducers (Eurosensors 2014), Brescia, Italy, 7–10 September 2014; pp. 456–459.
119. Miyamoto, K.; Seki, K.; Guo, Y.; Wagner, T.; Schöning, M.J.; Yoshinobu, T. Enhancement of the spatial resolution of the chemical imaging sensor by a hybrid fiber-optic illumination. In Proceedings of the 28th European Conference on Solid-State Transducers (Eurosensors 2014), Brescia, Italy, 7–10 September 2014; pp. 612–615.
120. Miyamoto, K.; Seki, K.; Suto, T.; Werner, C.F.; Wagner, T.; Schöning, M.J.; Yoshinobu, T. Improved spatial resolution of the chemical imaging sensor with a hybrid illumination that suppresses lateral diffusion of photocarriers. *Sens. Actuators B Chem.* **2018**, *273*, 1328–1333. [CrossRef]
121. Miyamoto, K.; Sakakita, S.; Wagner, T.; Schöning, M.J.; Yoshinobu, T. Application of chemical imaging sensor to in-situ pH imaging in the vicinity of a corroding metal surface. *Electrochim. Acta* **2015**, *183*, 137–142. [CrossRef]
122. Das, A.; Chen, T.C.; Lin, Y.T.; Lai, C.S.; Yang, C.M. Ultra-high scanning speed chemical image sensor based on light addressable potentiometric sensor with analog micro-mirror. In Proceedings of the SENSORS, 2013 IEEE, Baltimore, MD, USA, 3–6 November 2013; pp. 1412–1415.
123. Das, A.; Chen, T.C.; Yang, C.M.; Lai, C.S. A high-speed, flexible-scanning chemical imaging system using a light-addressable potentiometric sensor integrated with an analog micromirror. *Sens. Actuators B Chem.* **2014**, *198*, 225–232. [CrossRef]
124. Das, A.; Yang, C.M.; Chen, T.C.; Lai, C.S. Analog micromirror-LAPS for chemical imaging and zoom-in application. *Vacuum* **2015**, *118*, 161–166. [CrossRef]

125. Wagner, T.; Werner, C.F.; Miyamoto, K.; Schöning, M.J.; Yoshinobu, T. Development and characterisation of a compact light-addressable potentiometric sensor (LAPS) based on the digital light processing (DLP) technology for flexible chemical imaging. *Sens. Actuators B Chem.* **2012**, *170*, 34–39. [CrossRef]
126. Wagner, T.; Miyamoto, K.; Werner, C.F.; Schöning, M.J.; Yoshinobu, T. Utilising digital micro-mirror device (DMD) as scanning light source for light-addressable potentiometric sensors (LAPS). *Sens. Lett.* **2011**, *9*, 812–815. [CrossRef]
127. Lin, Y.H.; Das, A.; Lai, C.S. A simple and convenient set-up of light addressable potentiometric sensors (LAPS) for chemical imaging using a commercially available projector as a light source. *Int. J. Electrochem. Sci.* **2013**, *8*, 7062–7074.
128. Miyamoto, K.; Kaneko, K.; Matsuo, A.; Wagner, T.; Kanoh, S.; Schöning, M.J.; Yoshinobu, T. Miniaturized chemical imaging sensor system using an OLED display panel. *Sens. Actuators B Chem.* **2012**, *170*, 82–87. [CrossRef]
129. Werner, C.F.; Wagner, T.; Miyamoto, K.; Yoshinobu, T.; Schöning, M.J. High speed and high resolution chemical imaging based on a new type of OLED-LAPS set-up. *Sens. Actuators B Chem.* **2012**, *175*, 118–122. [CrossRef]
130. Miyamoto, K.; Kuwabara, Y.; Kanoh, S.; Yoshinobu, T.; Wagner, T.; Schöning, M.J. Chemical image scanner based on FDM-LAPS. *Sens. Actuators B Chem.* **2009**, *137*, 533–538. [CrossRef]
131. Miyamoto, K.; Itabashi, A.; Wagner, T.; Schöning, M.J.; Yoshinobu, T. High-speed chemical imaging inside a microfluidic channel. *Sens. Actuators B Chem.* **2014**, *194*, 521–527. [CrossRef]
132. Qintao, Z.; Ping, W.; Parak, W.J.; George, M.; Zhang, G. A novel design of multi-light LAPS based on digital compensation of frequency domain. *Sens. Actuators B Chem.* **2001**, *73*, 152–156. [CrossRef]
133. Wagner, T.; Werner, C.F.B.; Miyamoto, K.I.; Schöning, M.J.; Yoshinobu, T. A high-density multi-point LAPS set-up using a VCSEL array and FPGA control. *Sens. Actuators B Chem.* **2011**, *154*, 124–128. [CrossRef]
134. Itabashi, A.; Kosaka, N.; Miyamoto, K.; Wagner, T.; Schöning, M.J.; Yoshinobu, T. High-speed chemical imaging system based on front-side-illuminated LAPS. *Sens. Actuators B Chem.* **2013**, *182*, 315–321. [CrossRef]
135. Miyamoto, K.; Hirayama, Y.; Wagner, T.; Schöning, M.J.; Yoshinobu, T. Visualization of enzymatic reaction in a microfluidic channel using chemical imaging sensor. *Electrochim. Acta* **2013**, *113*, 768–772. [CrossRef]
136. Bousse, L.; Mcreynolds, R.J.; Kirk, G.; Dawes, T.; Lam, P.; Bemiss, W.R.; Farce, J.W. Micromachined multichannel systems for the measurement of cellular-metabolism. *Sens. Actuators B Chem.* **1994**, *20*, 145–150. [CrossRef]
137. Bousse, L.J.; Parce, J.W.; Owicki, J.; Kercso, K. Silicon micromachining in the fabrication of biosensors using living cells. In Proceedings of the IEEE 4th Technical Digest on Solid-State Sensor and Actuator Workshop, Hilton Head Island, SC, USA, 4–7 June 1990; pp. 173–176.
138. Miyamoto, K.I.; Sato, T.; Abe, M.; Wagner, T.; Schöning, M.J.; Yoshinobu, T. Light-addressable potentiometric sensor as a sensing element in plug-based microfluidic devices. *Micromachines* **2016**, *7*, 111. [CrossRef] [PubMed]
139. Liang, T.; Gu, C.; Gan, Y.; Wu, Q.; He, C.; Tu, J.; Pan, Y.; Qiu, Y.; Kong, L.; Wan, H. Microfluidic chip system integrated with light addressable potentiometric sensor (LAPS) for real-time extracellular acidification detection. *Sens. Actuators B Chem.* **2019**, *301*, 127004. [CrossRef]
140. Wu, J.; Chen, Q.; Lin, J.M. Microfluidic technologies in cell isolation and analysis for biomedical applications. *Analyst* **2017**, *142*, 421–441. [CrossRef]

Article

CCD Multi-Ion Image Sensor with Four 128 × 128 Pixels Array

Toshiaki Hattori [1,*], Fumihiro Dasai [1], Hikaru Sato [1], Ryo Kato [2] and Kazuaki Sawada [1,*]

1. Department of Electrical and Electronic Information Engineering, Toyohashi University of Technology, Hibarigaoka 1-1, Tenpaku, Toyohashi 441-8580, Japan; zr854452@bf7.so-net.ne.jp (F.D.); hikarusato0824@docomo.ne.jp (H.S.)
2. Cooparative Research Facility Center, Toyohashi University of Technology, Hibarigaoka 1-1, Tenpaku, Toyohashi 441-8580, Japan; ryo_kato@crfc.tut.ac.jp

* Correspondence: thattori@ee.tut.ac.jp (T.H.); sawada@ee.tut.ac.jp (K.S.); Tel.: +81-532-44-6806 (T.H.); Tel.: +81-532-44-6739 (K.S.)

Received: 29 January 2019; Accepted: 25 March 2019; Published: 1 April 2019

Abstract: A semiconductor array pH image sensor consisting of four separated blocks was fabricated using charged coupled device (CCD) and complementary metal oxide semiconductor (CMOS) technologies. The sensing surface of one of the four blocks was Si_3N_4 and this block responded to H^+. The surfaces of the other three blocks were respectively covered with cation sensitive membranes, which were separately printed with plasticized poly (vinyl chloride) solutions including Na^+, K^+, and Ca^{2+} ionophores by using an ink-jet printing method. In addition, each block of the image sensor with 128 × 128 pixels could have a calibration curve generated in each independent measurement condition. The present sensor could measure the concentration image of four kinds of ions (H^+, K^+, Na^+, Ca^{2+}) simultaneously at 8.3 frames per second (fps) in separated regions on a chip.

Keywords: CCD ion sensor; multi-ion image; CMOS technology; ink-jet printing; bioactive cations

---

1. Introduction

Semiconductor ion sensors are suitable for visualizing local chemical species, because of the capabilities of miniaturization and integration of the sensing area. Many miniaturized sensing pixels based on ISFET were two-dimensionally lined up neatly as an array-imaging sensor [1–5], and a semiconductor-sensing device was two-dimensionally divided into many sensing areas by an addressable light such as LAPS [6–10]. Although almost 50 years has passed since the introduction of the ISFET by Bergveld in 1970 [11], there are still many challenges to be overcome beyond ISFETOLOGY including the REFET subject, in which he summarized the device and readout circuity reported in 2010 [12]. In particular, the array image sensor requires a dense, reliable and scalable ISFET array that enables massive parallelism and high throughput [13]. Recently, several array image sensors were reported including 32 × 32 (1k) pixels and 9.3 frame/s (fps) image sensor [14]; 512 × 576 (300k) and 375 fps image sensor [15]; 3600 × 3600 (13M) and 26 fps image sensor [16]. The number of pixels and fps is increasing, and the spatial resolution is also improving. However, a high operation stability of the array image sensor is required. Therefore, it is important for a reliable array image sensor to demonstrate the image consisting of whole pixels, which indicates the total quality of achievement of the image sensor.

We had developed a different type of semiconductor array pH sensor (liner 8 pixels) based on charged coupled device (CCD) and complementary metal oxide semiconductor (CMOS) technologies in 1999 [17], as an aim at developing the image sensor. According to the plan to finely visualize local chemical species, the CCD pH sensor has steadily progressed to two-dimensional pH image sensors; from the initial stage of 10 × 10 (100) pixels and 30 fps [18]; 32 × 32 (1k) and 5 fps pixels [19]; 128 ×

128 and 50 fps (16k) pixels [20]; to 1320 × 976 and 27.5 (1.5M) pixels [21]. The size of sensing pixels became small, and the spatial resolution increased; 100 × 100 μm^2 [18], 23.55 × 23.55 μm^2 [19], 12.1 × 12.1 μm^2 [20], 3.75 × 3.75 μm^2 [21], respectively. Each image that consisted of whole pixels was demonstrated by an investigation of the sensor response to pH change.

The CCD pH image sensor can also convert into several CCD ion image sensors using the surface coating of ion sensitive membranes. The image sensors monitored local concentrations of metal ions such as K^+ [22], and biogenic amines [23]. The K^+ image sensor was applied to a tissue stimulation response of a cultured hippocampal slice stimulated by glutamate [22]. The biogenic amine image sensor was applied to mast cells stimulation response of mast cells by compound 48/80 [23]. On the other hand, living cells have several ion pumps and ion channels for H^+, Na^+, K^+, Ca^{2+}, Cl^-, and so on, to maintain the homeostasis. Therefore, a minute investigation of cell dynamics is required to simultaneously monitor these concentration changes. As the previous studies, we fabricated the CCD multi-ion image sensor for Na^+-K^+ [24], K^+-Ca^{2+} [25]. The K^+-Ca^{2+} image sensor was applied to the Ca^{2+} release of cultured PC12 cells stimulated by acetylcholine [25]. The multi-ion image sensor was superior to monitor their concentration changes simultaneously for only one experiment.

The multi-image sensors were successfully demonstrated, however, further developments were required; increments of numbers of ions measured simultaneously, clear distinction and refinement of separated sensing regions. For instance, multi-ion-vision equipment that displays a high-resolution image of many ions simultaneously such as a color television is one of the ultimate goals. In this paper, we fabricated a new CCD multi-ion image sensor with four divided regions of 128 × 128 pixels array in order to accomplish the measurement of several ions simultaneously and the clearly separated sensing regions on the chip. Each region of the new image sensor can individually configure the V_{ref}. Thus, the sensor can monitor four-ion concentration changes simultaneously. The establishment of the simultaneous monitoring technique for four ions is important in the developing process in order to develop the future multi-ion-vision equipment. The CCD multi-ion image sensor was fabricated from a newly designed CCD pH image sensor with four divided regions of 128 × 128 sensing pixels. Thereafter, three regions were prepared with Na^+, K^+, and Ca^{2+} sensitive membranes. These sensitive membranes were constructed by an ink-jet printing method. The present sensor could simultaneously determine the changes of pH, Na^+, K^+, and Ca^{2+} concentrations in an intracellular concentration region.

2. Fabrication Process and Readout Procedure

2.1. Pixel Design and Sensor Chip

The sensor system using CMOS technology based on the previous block diagram of 128 × 128 sensor [14] was designed and fabricated. The main circuit design and pixel structure were the same. The present ion image sensor has four times more pixels than the 128 × 128 pixels sensor. Here, the basic sensing principle was briefly mentioned using Figure 1 with the cross-sectional view of a sensing pixel. The key is to be able to measure the surface membrane potential between the aqueous solution and the membrane or the Si_3N_4 film in Figure 1. Hydrogen ions adsorb and desorb onto the Si_3N_4, and affect the surface membrane potential in the sensing area. On the other hand, when Si_3N_4 was covered with a plasticized polyvinyl chloride (PVC) membrane including a cation selective ionophore, the cations adsorb and desorb onto the membrane, and change the surface membrane potential. Thus, their surface membranes depend on the concentration of the sensitive cations. The sensor chip has four terminals to process the charge. The input control gate (ICG) and transfer gate (TG) electrodes are used to control the potential level as the gating. The input diode (ID) and floating diffusion (FD) are pn junctions that supply the charge and detect the quantity of charge, respectively. With a decreasing ID potential, the charge exceeds over ICG, and is filled at the charge sensor area corresponding to the membrane potential, as shown in Figure 1a. With an increasing ID potential, the charge is spilled and holds, as shown in Figure 1b. With an increasing TG, the charge is transferred into FD, as shown in

Figure 1c. The charge process can be repeated, and the charge is accumulated without noise. Therefore, the sensitivity is increasing, which is an excellent advantage for the CCD sensor.

The structure of the sensor chip and the wiring on the package board is shown in Figure 2. In order to hold a compact size and use previous equipment, the sensor chip is divided into four blocks and each block has a timing generator, shift register, horizontal and vertical scanner, respectively. Each terminal with the same name except for CE is electrically connected on a package board.

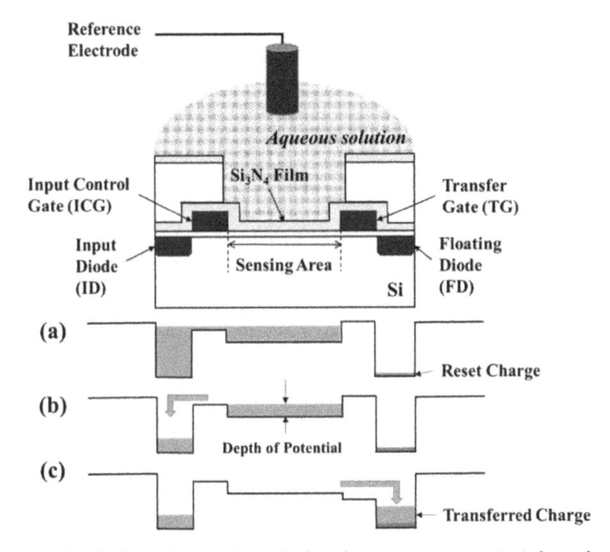

Figure 1. Cross-sectional view of a sensing pixel and measurement principle to focus on charges. (**a**) Charge supply into the depth of potential; (**b**) holding of charges corresponding to the depth; (**c**) transferring of charges to measure.

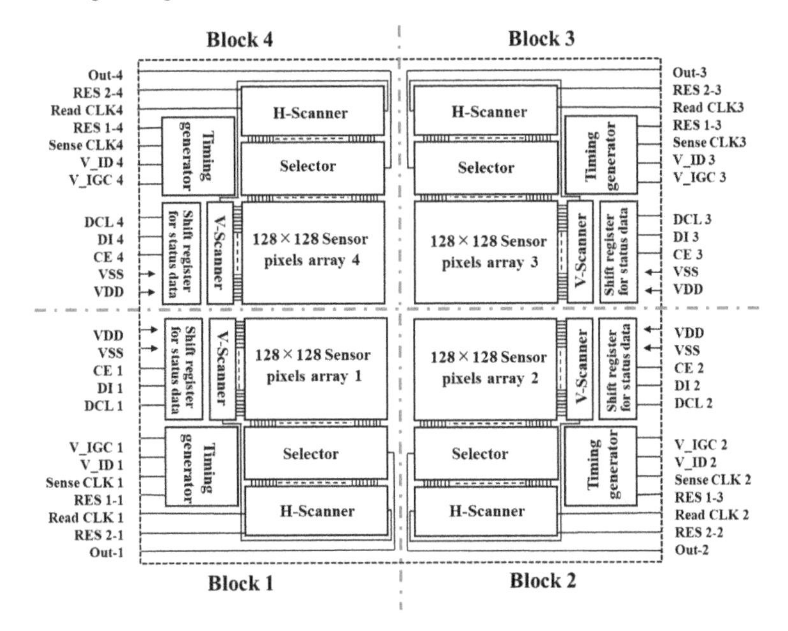

Figure 2. Block design for the present multi-ion sensor.

The switching time of each CE terminal is due to the order, as shown in Figure 3. The block whose CE terminal becomes high is activated. The measurement conditions data of each block are saved into Shift resister when CE terminal changes to be high.

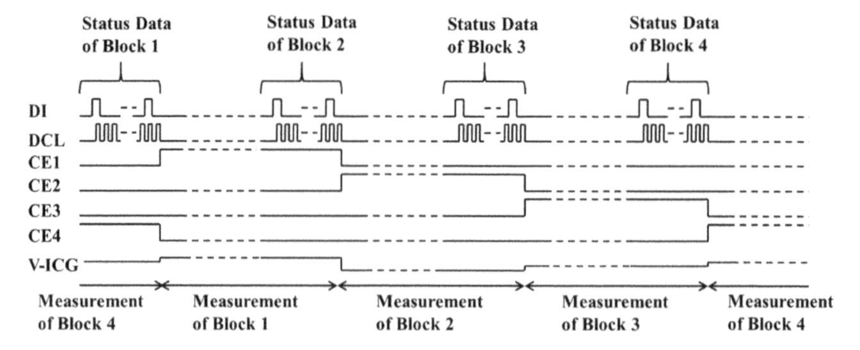

Figure 3. Serial timing chart for the operation of four blocks.

The whole photographs of a sensor chip and its extended image were shown in Figure 4. The 4 × 16K array CCD ion image sensor possessed 65,536 sensing pixels in a two-dimensional 4 × 128 × 128 array, as shown in Figure 4A. The dimensions of the pixel and the sensor area were 37.3 × 37.3 μm^2, 9.549 × 9.549 mm^2, respectively. The interval of blocks was 20 µm. The chip size was 12.2 × 12.2 mm^2. The frame speed was constant at 8.3 fps. The sensing area of a pixel was in a hollow of ~ 3 µm depth and its size was 13.5 µm × 24.5 µm, as shown in Figure 4B.

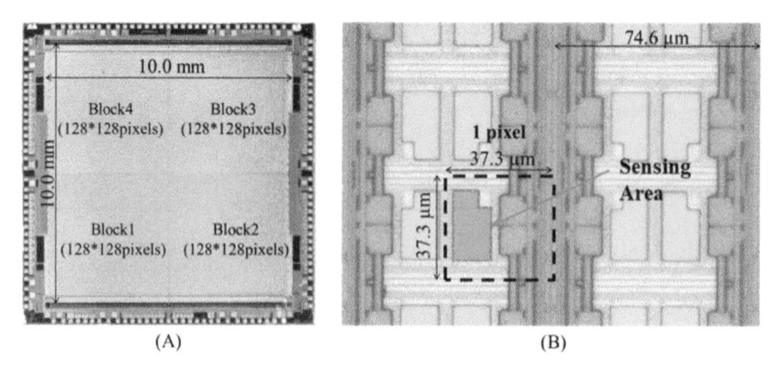

Figure 4. (**A**) A picture of the 256 × 256 array charged couple device (CCD)-type image sensor; (**B**) a micrograph of the sensor pixels.

2.2. Reagents

Bis[(12-crown-4)-methyl]-2-dodecyl-2-methylmalonate (bis(12-crown-4)), 2-Nitrophenyloctyl ether (NPOE), sodium tetrakis[3,5-bis(trifluoromethyl)phenyl]borate (NaTFPB) were purchased from Dojindo Lab., Inc. (Kumamoto, Japan). Polyvinylchloride (MW = 80,000) (PVC), 10,19-Bis [(octadecylcarbamoyl)methoxyacetyl]-1,4,7,13,16-pentaoxa-10,19-diazacycloheneicosane (K23E1), valinomycin, 1,3,5,7,9,11,13,15-oct(propylmethacryl)pentacyclo[9.5.1.13,9.15,15.17,13]octasiloxane (POSS), and tris(hydroxymethyl)aminomethane (Tris) were purchased from Sigma-Aldrich, Inc. (St. Louis, MO, USA). Potassium tetrakis(4-chlorophenyl)borate (K-TCPB) was purchased from Tokyo Chemical Industry (Tokyo, Japan). Dioctyl sebacate (DOS), n-dioctyl phthalate (DOP), tetrahydrofuran (THF), cyclohexanone (CHN), cyclohexane, and other reagents were purchased from Wako Pure Chemical Industries, Ltd. (Osaka, Japan). Water deionized by a Milli-Q system was used in all the experiments. Metal ion solutions were prepared daily by dilution from a 1 M stock solution of each metal chloride.

2.3. Ink-Jet Device and the Printing Conditions

An ink-jet apparatus (IJK-200T, Microjet, Shiojiri, Japan) was set up in a large glove box. The nozzle diameter (IJHB-300) was 300 μm. The apparatus was a piezoelectric dot ink-jet printer, and can place the printing solution at the desired point on a chip. The state of the produced drops was monitored using a strobe light camera. The proper adjustment of two voltage pulses on the operating piezoelectric device formed one drop. The pulse conditions were adjusted daily; 80.0–85.0 V as the pulse voltages, 5.0–10.0 μS as the 1st pulse range, 3.0–10.0 μS as the interval pulse range, 3.0–10.0 μS as the 2nd pulse range. One drop of the cocktail solution covered about 36 (6×6) sensing pixels. The rate of line printing and the interval of the pitch decided in each ion sensitive membranes. The waiting time before the recoating was 10 s. After printing, the membrane was dried for 24 h at room temperature. Since the ink-jet apparatus had only a single injection head, when a multi-ion membrane was prepared, each solution was separately printed.

2.4. Procedure of the Preparation of Each Sensitive Membrane

In order to increase the adhesion of the plasticized PVC membrane on the chip, the ink-jet printing areas were pretreated with sodium hydroxide solution. A 0.1 mM sodium hydroxide solution laid on the ink-jet printing area for more than 30 min, and washed with pure water, and the sensor was dried in a clean glove box at room temperature. After the pretreatment, each membrane was formed by the ink-jet method. For the Na^+ sensitive printing solution, PVC 44.0 mg, POSS 30.0 mg, DOP 22.0 mg, Bis (12-crown-4) 2.7 mg, TFPB 1.3 mg were dissolved in a solvent mixture of 5 mL of THF and 5 mL of CHN (THF-CHN solution). The mixed printing solution was dropped into the Na^+ sensing area with a number of overcoats of 20 times. At this time, the pitch of the drop was set to 100 μm. For the K^+ sensitive printing solution, PVC 30.0 mg, POSS 30.0 mg, DOS 36.0 mg, valinomycin 2.7 mg, K-TCPB 1.3 mg were dissolved in THF-CHN solution of 10 mL. The mixed printing solution was dropped into the K^+ sensing area with a number of overcoats of 20 times, and the pitch of the drop was set to 80 μm. For the Ca^{2+} sensitive printing solution, PVC 30.0 mg, POSS 11.2 mg, NPOE 55.2 mg, K23E1 3.0 mg, TFPB 1.3 mg were dissolved in THF-CHN solution of 10 mL. The mixed printing solution was dropped into the Ca^{2+} sensing area with the number of overcoats of 40 times, and the pitch of the drop was set to 80 μm.

2.5. Measurement of Ion Concentration

The electrochemical cell consisted of a LF-1/2 leak-less Ag/AgCl reference electrode sample solution/plasticized PVC membrane/semiconductor sensor. The potential slopes of sodium ion brock, potassium ion block, and calcium ion block were evaluated from three different concentrations of each ion solutions: 0.1 mol/L (M) Tris-buffered (pH 7.4) solutions with concentrations of 200, 20, and 2 mM sodium ions. 0.1 mol/L (M) Tris-buffered (pH 7.4) solutions with concentrations of 40, 4, and 0.4 mM potassium ions. 0.1 mol/L (M) Tris-buffered (pH 7.4) solutions with concentrations of 18, 1.8, and 0.18 mM sodium ions. These solutions with the calibration were corresponding to the range of extra-cellular concentration [26]. All signals of the sensing pixels were recorded as a video after each pixel arrayed on the chip was calibrated with the above three solutions.

3. Results and Discussion

3.1. Evaluation of Potential Response to Each Cation

The potential response of each block to the corresponding cation was shown in Figure 5a–d. To measure the hydrogen ion sensitivity of the sensor, ordinal buffered pH standard solutions (pH 4, pH 7, and pH 9) were used. The numbers of the right upper block to pH sensing pixels was 16,384.

The potential slope showed a distribution with an ideal high kurtosis. The maximum of the frequency distribution of the pH sensitivity was 46 mV/pH (48 mV/pH to pH 4–7, 43 mV/pH to pH 7–9). Although the slope is inferior to Nernstian response, most of the slopes indicated were highly sensitive. The other three blocks were treated with the ink-jet printing. The left upper block of the chip was printed with the sodium ion membrane, and the left under block with the calcium ion membrane, the right under block with the potassium ion membrane. The potential slope of sodium ion block is high and the maximum distribution was 61 [mV/decade] (63 mV/decade to 0.4 mM–4 mM, 60 mV/decade to 4 mM–40 mM) that were almost Nernstian response, as shown in Figure 5b. The frequency distribution has a high kurtosis with a deviation to high value. The potential slope of the potassium ion block is lower than that of the sodium ion block, as shown in Figure 5c. The maximum distribution potential slope was 52 [mV/decade] (53 mV/decade to 2 mM–20 mM, 51 mV/decade to pH 20 mM–200 mM), and has a wide normal distribution. Although many potential responses did not achieve Nernstian slope, the sensitivity was higher than the pH block. The potential slope of calcium ion block is lower than the other cation, as shown in Figure 5d. The lower sensitivity was due to the ion valence of +2. The maximum distribution potential slope was 29.0 [mV/decade] (26 mV/decade to 0.18 mM–1.8 mM, 32 mV/decade to 1.8 mM–18 mM) that was Nernstian slope. The frequency distribution has a high kurtosis and a slight distribution with a small deviation to high value. These potential slopes were not ideal, however, the response was close to Nernstian response or sufficient response.

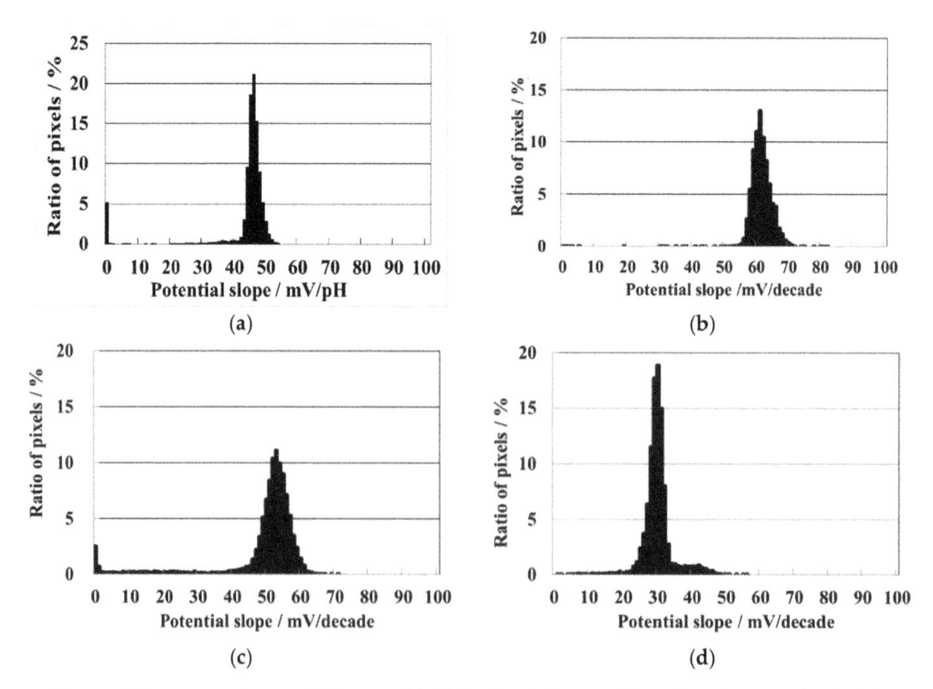

Figure 5. Histograms of potential slopes to (**a**) pH for the rigth upper block; (**b**) sodium ion for the left upper block; (**c**) potassium ion for the right under block; (**d**) calcium ion for the left under block.

3.2. Dynamic Response

Snapshots of the multi-ion image sensor are shown in Figure 6. When 400 μL of the 200 mM sodium ion solution was added into 400 μL of 2 mM sodium ion solution by a 1000 μL pipette, the color of the left upper block immediately changed from blue to red (Figure 6a). The left upper block responded to the concentration change of sodium ions, but no change appeared in the other block, except for slight areas at the upper edge of calcium ion block and the right edge of potassium ion block. When the sodium membrane was prepared by the ink-jet method, it seemed that the sodium membrane painting solution was leaked to these areas. When 400 μL of the 0.18 mM calcium ion solution was added into 400 μL of 18 mM calcium ion solution by a 1000 μL pipette, the color of the left under block immediately changed from blue to red (Figure 6b). The left under block responded to the concentration change of calcium ions, but no change appeared in the other block. When 400 μL of the 0.4 mM potassium ion solution was added into 400 μL of 40 mM potassium ion solution by a 1000 μL pipette, the color of the right under block immediately changed from blue to red (Figure 6c). The right under block responded to the concentration change of potassium ions, but no change appeared in the other block, except for the sodium ion block. The sodium ion block had a slightly potential response to calcium ion concentration. All the concentration changes were completed in three s. For all additions of the pipette, the pH region of the upper right did not show a concentration change. This is very effective to selectively detect ion behaviors based on cell functions without considering pH variation.

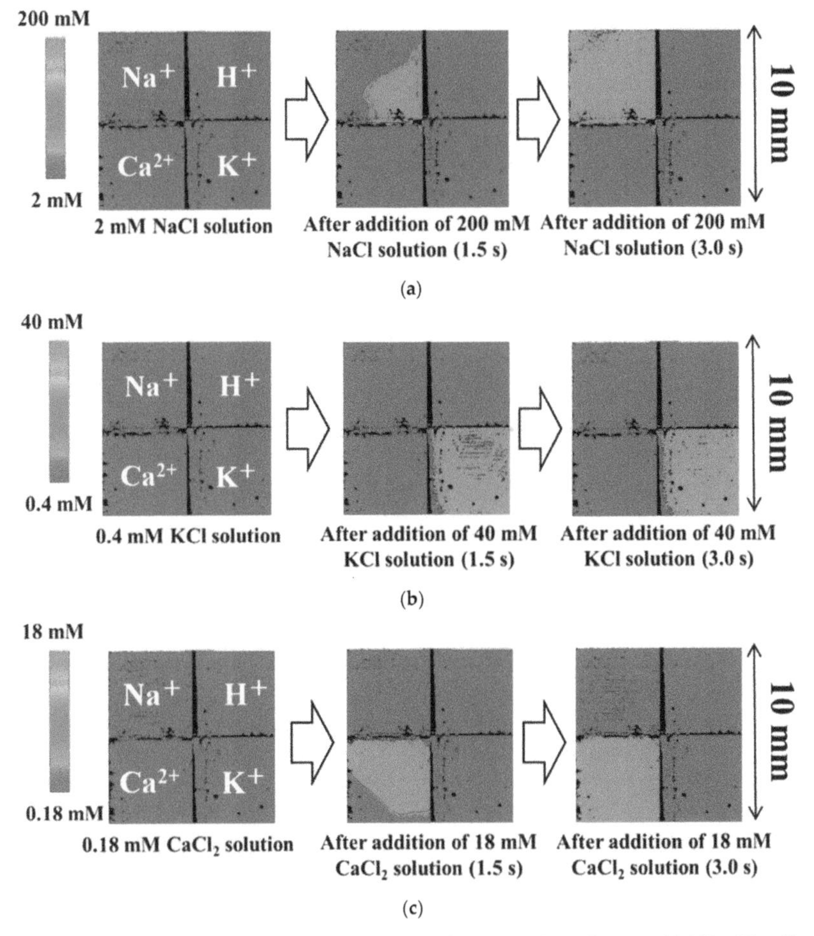

Figure 6. Snapshots of CCD multi-ion image sensor with 128 × 128 pixels array. (**a**) 200 mM sodium ion solution (100 μL) was injected into a 400 μL of sample solution including 2 mM sodium ion solution by a pipette; (**b**) 40 mM potassium ion solution (100 μL) was injected into a 400 μL of sample solution including 0.4 mM potassium ion solution by a pipette; (**c**) 18 mM calcium ion solution (100 μL) was injected into a 400 μL of sample solution including 0.18 mM calcium ion solution by a pipette.

In order to demonstrate the visualization of the mixture solution, a phosphate buffer solution of 100 μL (pH 6.86, 0.025 M di-hydrogen potassium phosphate +0.025 M hydrogen di-sodium phosphate) was added to a borate buffer solution of 400 μL (pH 9.18, 0.01 M borate solution of half-neutralization by NaOH), as shown in Figure 7. The color of the Na^+ region gradually changed because of increasing Na^+ concertation from 0.005 M to 0.01 M. The color of the H^+ quickly changed because of pH 9 to about 7. The color of part of the K^+ region changed because of increasing K^+ concentration to 0.005 M. Unfortunately, the preparation of K^+ membrane of this chip was insufficient. Since the two solutions did not contain Ca^{2+}, no color change was observed in the Ca^{2+} region. The change of these colors was reasonable so that the simultaneous visualization by the change of three ions was demonstrated.

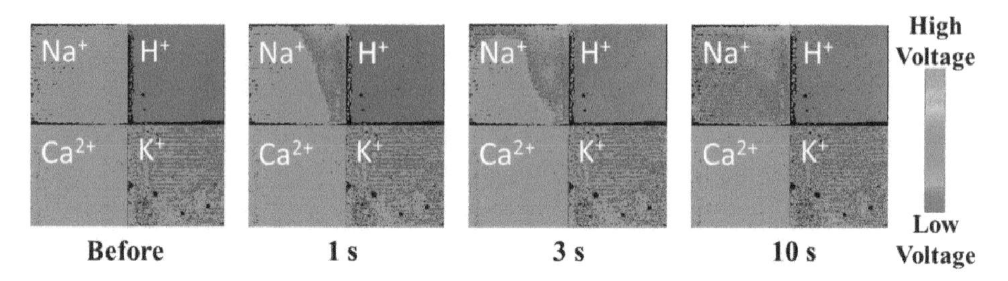

Figure 7. Visual image changes for the instant addition of a phosphate buffer solution (pH 7) to a borate buffer solution (pH 9). This sensor chip is different from the sensor chip shown in Figure 6.

4. Conclusions

We fabricated a CCD ion multi-image sensor that has a sodium ion region, a potassium ion region, a calcium ion region, and a hydrogen ion region. Each region except for the hydrogen ion region was respectively covered with the cation sensitive membrane, which was formed on the chip by an ink-jet method. Each sensitive block sufficiently responded to the corresponding cation. In addition, we were able to observe a dynamic change of ion concentration as an animation. These results showed that this sensor didn't have any problem as the structure of the multi-ion sensor essentially. This type of sensor will be one of the useful tools for biochemical application. The subjects of increments of numbers of ions measured simultaneously, and the distinction of separated sensing regions was accomplished. The concentration of the measurable cations is relatively high at the biological environment. Therefore, the sensitivity, which is the ability to identify the difference between large concentrations, rather than the detection limit is required. The present sensor can demonstrate the difference of the Ca^{2+} release of cultured PC12 cells stimulated by acetylcholine at least, because of the same basic design as the Ca^{2+}-K^+ sensor [25]. Moreover, the increment of accumulation measurable cycles by a function which was furnished in the present sensor system can further reduce the noise and improve the sensitivity to be able to distinct the concentration of less than 0.01 of pH, or −log [M] [27]. Except for milli-sec order study such as neural cells, sec order change of cells and tissue can be visualized. However, our CCD sensors are still in the developing stage for the future multi-ion-vision equipment, four blocks were large on the present sensor. Furthermore, refinement of sensing regions is required. Therefore, we are now planning to develop the multi-ion sensor which has small (<30 μm) and numerous ion sensing regions which are uniformly placed in the whole chip. On the other hand, improvement of the frame rate of the imaging is also required. The present sensor system has only one single analog to digital converter (ADC), so the frame rate is reduced by increasing the numbers of sensing pixels. Therefore, we will be able to improve the frame rate by increasing with the number of ADC.

Author Contributions: Conceptualization, T.H., F.D. and K.S.; methodology, F.D.; software, T.H.; validation, T.H., F.D. and K.S.; formal analysis, T.H. and H.S.; investigation, H.S. and R.K.; resources, T.H.; data curation, T.H.; writing—original draft preparation, T.H.; writing—review and editing, T.H. and F.D.; visualization, T.H.; supervision, K.S.; project administration, K.S.; funding acquisition, K.S.

Funding: This work was supported by JST CREST Grant Number JPMJCR14G2, Japan.

Conflicts of Interest: The authors declare no conflict of interest.

References

1. Meyer, H.; Drewer, H.; Krause, J.; Cammann, K.; Kakerow, R.; Manoli, Y.; Mokwa, W.; Rosrert, M. Chemical and biochemical sensor array for two-dimensional imaging of anlyte distributions. *Sens. Actuators B Chem.* **1994**, *18*, 229–234. [CrossRef]
2. Kakerow, R.; Manoli, Y.; Moka, W.; Rosrert, M.; Rospert, M.; Meyer, H.; Drewer, H.; Krause, J.; Cammann, K. A monolithic sensor array of individually addressable microelectrodes. *Sens. Actuators A Phys.* **1994**, *43*, 296–301. [CrossRef]
3. Nakazato, K. An integrated ISFET sensor array. *Sensors* **2009**, *9*, 8831–8851. [CrossRef]
4. Nemeth, B.; Piechocinski, M.S.; Cumming, D.R.S. High-resolution real-time ion-camera system using a CMOS-based chemical sensor array for proton imaging. *Sens. Actuators B Chem.* **2012**, *171–172*, 747–752. [CrossRef]
5. Machida, S.; Shimada, H.; Motoyama, Y. Multiple-channel detection of cellular activities by ion-sensitive transistors. *Jpn. J. Appl. Phys.* **2018**, *57*, 04FM03. [CrossRef]
6. Hafeman, D.G.; Parce, J.W.; McConnell, H.M. Light-addressable potentiometric sensor for biochemical systems. *Science* **1988**, *240*, 1182. [CrossRef] [PubMed]
7. Qintao, Z.; Ping, W.; Parak, W.J.; George, M.; Zhang, G. A novel design of multi-light LAPS based on digital compensation of frequency domain. *Sens Actuators B Chem.* **2001**, *73*, 152–156. [CrossRef]
8. Yoshinobu, T.; Schöning, M.J.; Finger, F.; Moritz, W.; Iwasaki, H. Fabrication of thin-film LAPS with amorphous silicon. *Sensors* **2004**, *4*, 163–169. [CrossRef]
9. Guo, Y.; Seki, K.; Miyamoto, K.; Wagner, T.; Schöning, M.J.; Yoshinobu, T. Novel photoexcitation method for light-addressable potentiometric sensor with higher spatial resolution. *Appl. Phys. Exp.* **2014**, *7*, 067301. [CrossRef]
10. Wang, J.; Du, L.; Krause, S.; Wu, C.; Wang, P. Surface modification and construction of LAPS towards biosensing applications. *Sens. Actuators B Chem.* **2018**, *265*, 161–173. [CrossRef]
11. Bergveld, P. Development of an ion-sensitive solid-state device for neurophysiological measurements. *IEEE Trans. Biomed. Eng.* **1970**, *17*, 70. [CrossRef]
12. Bergveld, P. Thirty years of ISFETOLOGY: What happened in the past 30 years and what may happen in the next 30 years. *Sens Actuators B Chem.* **2003**, *88*, 1–20. [CrossRef]
13. Moser, N.; Lande, T.S.; Toumazou, C.; Georgiou, P. ISFETs in CMOS and emergent trends in instrumentation: A review. *IEEE Sens. J.* **2016**, *16*, 6496–6514. [CrossRef]
14. Hu, Y.; Moser, N.; Georgiou, P. A 32 × 32 ISFET Chemical Sensing Array With Integrated Trapped Charge and Gain Compensation. *IEEE Sens. J.* **2017**, *17*, 5276–5284. [CrossRef]
15. Jiang, Y.; Liu, X.; Huang, X.; Guo, J.; Yan, M.; Yu, H.; Huang, J.-C.; Hsieh, C.-H.; Chen, T.-T. A 201 mV/pH, 375 fps and 512×576 CMOS ISFET sensor in 65nm CMOS technology. In Proceedings of the IEEE Custom Integrated Circuits Conference (CICC), San Jose, CA, USA, 28–30 September 2015; pp. 1–4.
16. Cong, Y.; Xu, M.; Zhao, D.; Wu, D. A 3600 × 3600 large-scale ISFET sensor array for high-throughput pH sensing. In Proceedings of the IEEE 12th International Conference on ASIC (ASICON), Guiyang, China, 25–28 October 2017; pp. 957–960.
17. Sawada, K.; Mimura, S.; Tomita, K.; Nakanishi, T.; Tanabe, H.; Ishida, M.; Ando, T. Novel CCD-based pH imaging sensor. *IEEE Trans. Electron. Devices* **1999**, *46*, 1846–1849. [CrossRef]
18. Hizawa, T.; Sawada, K.; Takao, H.; Ishida, M. Fabrication of a two-dimensional pH image sensor using a charge transfer technique. *Sens. Actuators B Chem.* **2006**, *117*, 509–515. [CrossRef]
19. Hizawa, T.; Matsuo, J.; Ishida, T.; Takao, H.; Abe, H.; Sawada, K.; Ishida, M. 32 × 32 pH image sensors for real time observation of biochemical phenomena. In Proceedings of the International Solid-State Sensors, Actuators and Microsystems Conference, Lyon, France, 10–14 June 2007; pp. 1311–1312.
20. Futagawa, M.; Suzuki, D.; Otake, R.; Dasai, F.; Ishida, M.; Sawada, K. Fabrication of a 128 × 128 Pixels Charge Transfer Type Hydrogen Ion Image Sensor. *IEEE Trans. Electron. Dev.* **2013**, *60*, 2634–2639. [CrossRef]

21. Edo, Y.; Tamai, Y.; Yamazaki, S.; Inoue, Y.; Kanazawa, Y.; Nakashima, Y.; Yoshida, T.; Arakawa, T.; Saitoh, S.; Maegawa, M.; et al. 1.3 Mega pixels CCD pH imaging sensor with 3.75 μm spatial resolution. In Proceedings of the IEEE International Electron Devices Meeting (IEDM), Washington, DC, USA, 7–9 December 2015; pp. 29.3.1–29.3.4.
22. Kono, A.; Sakurai, T.; Hattori, T.; Okumura, K.; Ishida, M.; Sawada, K. Label free bio image sensor for real time monitoring of potassium ion released from hippocampal slices. *Sens. Actuators B Chem.* **2014**, *201*, 439–443. [CrossRef]
23. Hattori, T.; Tamamura, Y.; Tokunaga, K.; Sakurai, T.; Kato, R.; Sawada, K. Two-dimensional microchemical observation of mast cell biogenic amine release as monitored by a 128 × 128 array-type charge-coupled device ion image sensor. *Anal. Chem.* **2014**, *86*, 4196–4201. [CrossRef]
24. Hattori, T.; Satou, H.; Tokunaga, K.; Kato, R.; Sawada, K. 16K Array Charge Coupled Device Multi-Ion Image Sensors for Simultaneous Determination of Distributions of Sodium and Potassium Ions. *Sens. Mater.* **2015**, *27*, 1023–1034.
25. Matsuba, S.; Kato, R.; Okumura, K.; Sawada, K.; Hattori, T. Extracellular Bio-imaging of Acetylcholine-stimulated PC12 Cells Using a Calcium and Potassium Multi-ion Image Sensor. *Anal. Sci.* **2018**, *34*, 553–558. [CrossRef] [PubMed]
26. Lodish, H.; Berk, A.; Kaiser, C.A.; Krieger, M.; Bretscher, A.; Ploegh, H.; Amon, A.; Martin, K. *Molecular Biology of the Cells*, 8th ed.; W. H. Freemann and Company: New York, NY, USA, 2016; p. 485.
27. Watanabe, E.; Hizawa, T.; Mimura, T.; Ishida, T.; Takao, K.; Sawada, K.; Ishida, M. Low-noise operation of charge-transfer-type ph sensor using charge accumulation technique. In Proceedings of the 11th International Conference on Miniaturized Systems for Chemistry and Life Sciences, Paris, France, 7–11 October 2007; pp. 479–481.

Article

InGaN as a Substrate for AC Photoelectrochemical Imaging

Bo Zhou [1], Anirban Das [1], Menno J. Kappers [2], Rachel A. Oliver [2], Colin J. Humphreys [1] and Steffi Krause [1,*]

[1] School of Engineering and Materials Science, Queen Mary University of London, Mile End Road, London E1 4NS, UK; b.zhou@qmul.ac.uk (B.Z.); a.das@qmul.ac.uk (A.D.); c.humphreys@qmul.ac.uk (C.J.H.)
[2] Department of Materials Science and Metallurgy, University of Cambridge, 27 Charles Babbage Road, Cambridge CB3 0FS, UK; mjk30@cam.ac.uk (M.J.K.); rao28@cam.ac.uk (R.A.O.)
* Correspondence: s.krause@qmul.ac.uk; Tel.: +44-2078823747

Received: 16 August 2019; Accepted: 8 October 2019; Published: 11 October 2019

Abstract: AC photoelectrochemical imaging at electrolyte–semiconductor interfaces provides spatially resolved information such as surface potentials, ion concentrations and electrical impedance. In this work, thin films of InGaN/GaN were used successfully for AC photoelectrochemical imaging, and experimentally shown to generate a considerable photocurrent under illumination with a 405 nm modulated diode laser at comparatively high frequencies and low applied DC potentials, making this a promising substrate for bioimaging applications. Linear sweep voltammetry showed negligible dark currents. The imaging capabilities of the sensor substrate were demonstrated with a model system and showed a lateral resolution of 7 microns.

Keywords: photoelectrochemistry; InGaN/GaN epilayer; cell imaging; light-activated electrochemistry; light-addressable potentiometric sensor

---

1. Introduction

Over the past three decades since first being proposed by Hafeman et al. in 1988 [1], photocurrent imaging with light-addressable potentiometric sensors (LAPS) has received increasing attention for chemical and biological applications such as the detection of ions [2], redox potentials [3], enzymatic reactions [4] and cellular activities [5–7]. By scanning a designated area of an electrolyte–insulator–semiconductor (EIS) structure with a modulated light beam, spatiotemporal AC photocurrent images with the two-dimensional distribution of analytes are produced [8,9].

To enhance the spatial resolution and photocurrent response, a wide range of semiconductor substrates have been investigated. Silicon on insulator (SOI) [10,11], ultrathin silicon on sapphire (SOS) [12] and semiconductor materials such as amorphous silicon, GaAs [13], GaN [14], TiO_2 [15] and In-Ga-Zn oxide [16] have been studied. SOS substrates exhibited a high resolution of 1.5 μm with a focused 405 nm laser beam and a resolution of 0.8 μm using a two-photon effect with a 1250 nm femtosecond laser [12]. SOS functionalized with self-assembled monolayers (SAMs) as an insulator has been used for imaging of chemical patterns [17–19], microcapsules [20], and yeast cells [21]. Modifying silicon with SAMs terminated with redox active species allowed the imaging of photo-induced redox currents [22].

Recently, ITO-coated glass without any insulator was proposed as a low-cost and robust substrate for photoelectrochemical imaging [23,24]. In the absence of an insulator, the AC photocurrent is largely determined by the anodic oxidation of hydroxide making ITO-LAPS highly sensitive to pH (70 mV/pH). Photocurrent imaging with ITO-LAPS showed a good lateral resolution of 2.3 μm [23] and was confirmed to be sensitive to the surface charge of living cells [24]. ZnO nanorods were used as a substrate for AC photocurrent imaging to monitor the degradation of a thin poly (ester amide) film with the enzyme α-chymotrypsin, also showing great potential in biosensing and bioimaging applications [25]. However, a relatively high applied bias (1.5 V) was required to achieve sufficiently high photocurrents with ITO and ZnO nanorods for two-dimensional imaging, which could possibly interfere with cellular metabolism. Moreover, due to low charge carrier mobility, both ITO and ZnO suffered a dramatic decrease in photocurrent with increasing modulation frequency, resulting in a low working frequency of 10 Hz for imaging. This could consequently limit their application for high-speed imaging, which is required for the investigation of cellular responses.

In this work, InGaN/GaN on sapphire was investigated as a new substrate for AC photoelectrochemical imaging, aiming to solve the above-mentioned problems. InGaN is a semiconductor alloy with a direct band gap that can be tuned from the near-infrared (0.6 eV, InN) to the ultraviolet (3.4 eV, GaN) by adjusting the indium concentration. It has been used widely in developing LEDs [26,27] and photovoltaic devices [28] owing to its strong light emission and absorption and a wide range of band gaps. InGaN has also gained significant attention in photoelectrochemistry. With band edges straddling oxygen and hydrogen redox overpotentials, p-type GaN/InGaN nanowires have been investigated in water splitting [29], having the advantages of high carrier mobility, good chemical stability and band gap tunability. GaN/InGaN nanowires have also been shown to exhibit excellent optochemical and electrochemical sensor performance, achieving the detection of pH [30], oxidizing gases (O_2, NO_2 and O_3) [31] through photoluminescence, and electrochemical detection of nicotinamide adenine dinucleotide (NADH) [32]. In this work, it will be shown that epitaxial layers of InGaN are suitable for photoelectrochemical imaging with good lateral resolution and have great potential in bioimaging applications.

2. Experimental Section

2.1. Materials

The InGaN/GaN structure was grown on a two-side polished (0001) sapphire substrate in a Thomas Swan 6 × 2″ metalorganic vapor-phase epitaxy reactor using trimethyl gallium (TMG), trimethyl indium (TMI), silane (SiH_4) and ammonia (NH_3) as precursors, while purified hydrogen and nitrogen were used as the carrier gases. A 40-nm-thick low-temperature (580 °C) GaN nucleation layer was followed by a 100-nm-thick n-type GaN layer deposited at 1060 °C in a hydrogen atmosphere at a constant pressure of 100 Torr. The carrier gas was then switched to nitrogen, the pressure ramped at 300 Torr and the temperature to 770 °C for the growth of the 100-nm-thick n-type InGaN epilayer.

All wet chemicals were purchased from Sigma-Aldrich (Gillingham, UK). All solutions in this work were prepared using ultrapure water (18.2 MΩ cm) from a Milli-Q water purification system (Millipore, Burlington, MA, USA).

2.2. Preparation and Characterization of Sensor Chip

The InGaN/GaN structure was cut into 5 mm × 5 mm pieces. These were ultrasonically cleaned with acetone, isopropanol and ultrapure water each for 15 min and blow dried with nitrogen. The InGaN/GaN samples were kept at room temperature before use. The morphology of InGaN/GaN was examined using a scanning electron microscope (SEM, FEI Inspect F, Thermo Fisher Scientific, Hillsboro, OR, USA). Ultraviolet–visible (UV-vis) spectra were obtained using a UV-Vis spectrometer (Lamda 950, PerkinElmer, Seer Green, UK).

2.3. Linear Sweep Voltammetry (LSV)

LSV of InGaN/GaN was carried out in Dulbecco's Phosphate Buffered Saline (DPBS) solution (pH 7.4) using an Autolab PGSTAT30/FRA2 electrochemical workstation (Windsor Scientific Ltd., Slough, UK). A platinum electrode and an Ag/AgCl (3 M KCl) electrode were the counter electrode and reference electrode, respectively. The scan rate was 10 mV/s. A diode laser (λ = 405 nm, max 50 mW), chopped in 10 s intervals was used as the light source while recording the LSV curves.

2.4. Cell Culture

Before seeding cells, InGaN substrates were sterilized with 70% ethanol and rinsed thoroughly with sterilized DPBS solution and blown dry. MG-63 human osteosarcoma cells were cultivated in Dulbecco's Modified Eagle's Medium (DMEM, Cat No D6429) supplemented with 10% Fetal Bovine Serum (FBS, Cat No F9665) and 1% penicillin-streptomycin (Cat No P4333) in an air jacketed incubator with 5% CO_2 at 37 °C with the medium changed every two days. At 70–80% confluence, cells were trypsinized by using Trypsin-EDTA (Cat No T3924), and resuspended in 10% FBS-supplemented DMEM, seeded onto the InGaN surface at a concentration of 2.5×10^4 cells/mL and incubated at 37 °C with 5% CO_2 for 24 h.

The cell viability was tested using a fluorescence live/dead assay (Thermo Fisher Scientific, Hillsboro, OR, USA, cat. no.: L3224). MG-63 cells were seeded onto two pieces of InGaN (5 mm × 5 mm) assembled in the photoelectrochemical imaging chamber at a concentration of 9.4×10^5 cells/mL and incubated at 37 °C with 5% CO_2 for 24 h. One InGaN chip was subjected to a raster scan in DPBS while another stayed under ambient conditions for the same time. Then, 0.5 mL of 2 μM calcein AM, 4 μM Ethidium homodimer-1 and 8.12 μM of Hoechst 33342 was used to detect the viability of the cells with and without AC photoelectrochemical imaging. Three different areas in each sample were checked using a fluorescence microscope (Leica DMI4000B Epifluorescence, Leica Microsystems Ltd., Milton Keynes, UK), and cell photos were then processed by Image J software for counting cells.

2.5. AC Photocurrent Imaging

Figure 1 depicts the LAPS set-up used in this work. A diode laser LD1539 (Laser 2000, Huntingdon, UK, λ = 405 nm, max 50 mW) intensity modulated at 1 kHz was used as the light source. The sample chamber was mounted onto an M-VP-25XL XYZ positioning system with a 50 nm motion sensitivity on all axes (Newport, UK). AC photocurrents were measured with an EG&G 7260 lock-in amplifier with a platinum electrode and an Ag/AgCl (3 M KCl) electrode acting as the counter and reference electrodes, respectively. DPBS (pH 7.4) was used as the electrolyte. Optical images of the sensor surface were obtained with a CMOS camera by illuminating the chip surface with white light from the front side. A drop of poly(methyl methacrylate) (PMMA) was deposited on the InGaN surface and dried overnight to obtain a model system for measuring the resolution.

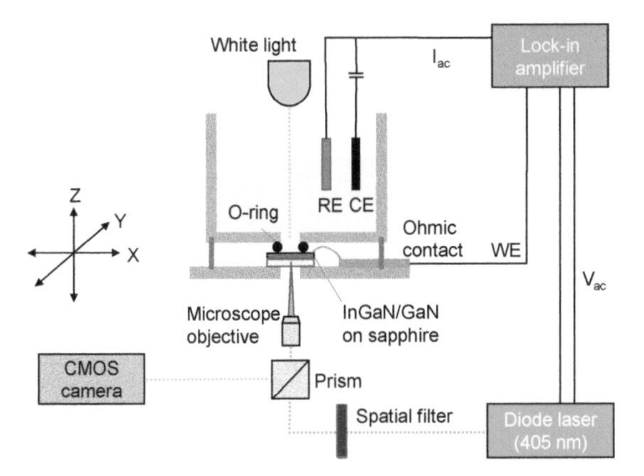

Figure 1. Schematic of the LAPS setup with a 405 nm diode laser to generate photo-induced charge carriers, a lock-in amplifier to measure AC photocurrent, and an X-Y-Z stage to move the electrochemical cell with respect to the laser beam for imaging.

3. Results and Discussion

3.1. Characterization of InGaN/GaN Epilayers on Sapphire

The SEM analysis of the InGaN/GaN structure is presented in Figure 2. The SEM top view in Figure 2a shows the InGaN surface with a high density of pits ((2.26 ± 0.08) × 10^{10} pits/cm^2) ranging between 20 nm and 50 nm in diameter, as some of the pits have merged. These "V-pits" are well known in InGaN growth and consist of an inverted hexagonal pyramid emanating from a threading dislocation formed at the sapphire/GaN interface. The pits open up during InGaN growth, which takes place at relatively low temperatures [33]. The total thickness of the InGaN/GaN epilayer was about 216.5 ± 6.6 nm, as shown in Figure 2b. Four-probe electrical measurements using soldered indium contacts showed a resistivity of 0.02 Ω·cm due to the n-type conductivity of the epilayers. A photoluminescence (PL) spectral map (Accent RPM2000, exc = 266 nm) of the 2-inch wafer showed a strong emission band centered at 448 ± 2 nm indicating an average indium fraction of ca. 17.5% [34].

Figure 3 shows the UV-Vis absorption spectrum of InGaN/GaN. From the inset Tauc-plot [35,36], a direct band gap of 2.77 ± 0.03 eV was determined, indicating that the charge carriers in InGaN/GaN are excited at wavelengths ≤ 448 nm, which is in good correspondence with the PL mapping result.

The DC photocurrent response of the InGaN/GaN sample was characterized with LSV. As shown in Figure 4, significant photocurrents were observed at anodic potentials ≥ 0 V. The dark current was negligible compared to the photocurrent. As with ITO substrates, the photocurrent can be ascribed to the oxidation of hydroxide ions in the solution. In contrast to ITO, the InGaN layers show a much lower onset potential of the photocurrent.

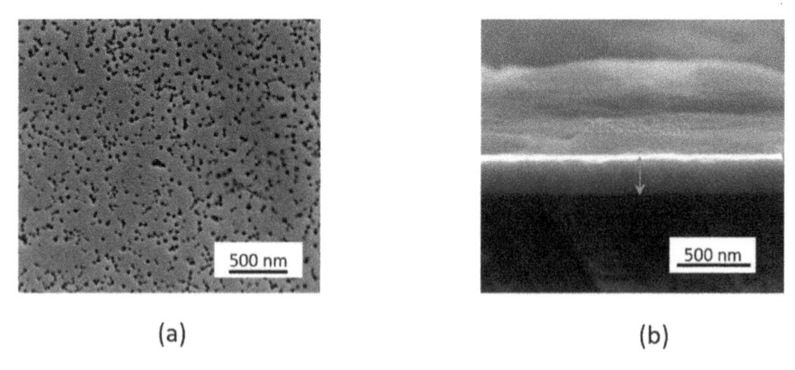

Figure 2. SEM images of InGaN/GaN: (**a**) top view and (**b**) cross-sectional view.

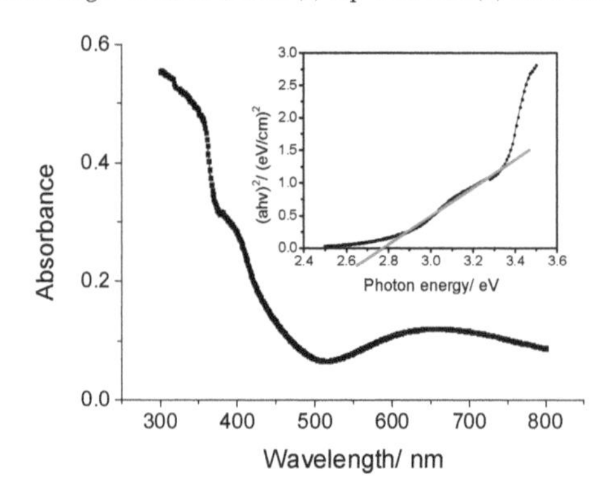

Figure 3. UV-Vis spectrum of InGaN and inset Tauc-plot.

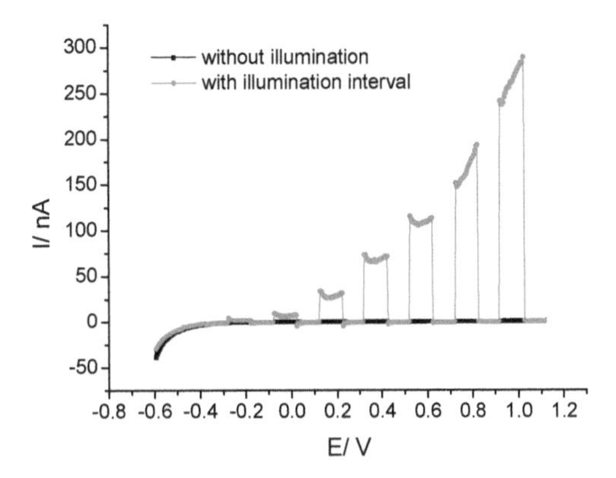

Figure 4. LSV curves of InGaN in the dark and with chopped illumination.

Figure 5a shows the dependence of the AC photocurrent on the modulation frequency measured at 1.0 V with a focused laser beam. From 10 Hz to 3 kHz, the photocurrent did not change significantly with the frequency, and then it decreased at higher frequencies. Significant photocurrents were obtained up to modulation frequencies of 10 kHz. The photocurrent became negligible at frequencies greater that 20 kHz. In contrast, the AC photocurrent measured with ITO and ZnO previously decreased continuously, with increasing modulation frequency above 10 Hz for ITO [23] and above 30 Hz for ZnO [25], becoming negligible at 7 kHz for ITO and 4 kHz for ZnO. This can be attributed to the significantly higher hole mobilities in InGaN [37] compared to those in ITO [38] and ZnO [39], as low-mobility minority charge carriers will not contribute to the AC photocurrent at high frequencies. In this work, 1 kHz was chosen as the modulation frequency since it could offer high quality images while also demonstrating the potential for high-speed imaging.

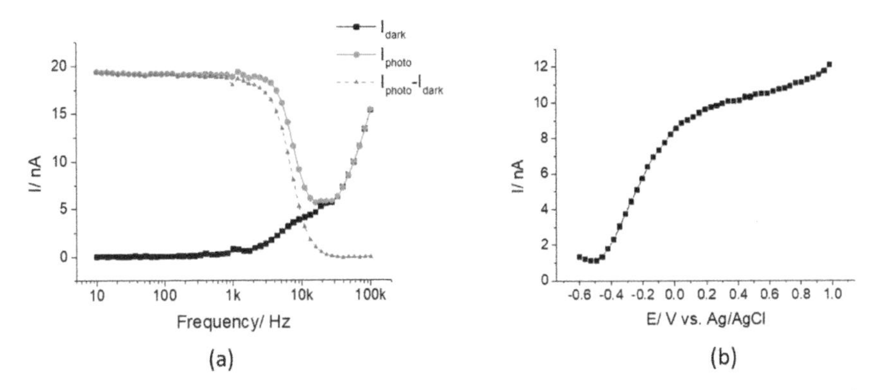

Figure 5. (**a**) Frequency dependence of the AC photocurrent and the background dark current measured at 1.0 V; (**b**) Characteristic $I-V$ curve of InGaN/GaN measured in pH 7.4 DPBS at 1 kHz with a focused laser beam at 18% maximum intensity.

Figure 5b shows the characteristic AC photocurrent–voltage ($I-V$) curve of InGaN/GaN in the voltage range -0.6 V to 1.0 V in pH 7.4 DPBS under the illumination of a focused laser beam (modulation frequency was 1 kHz). It shows that the photocurrent increased with the applied bias, to a value of 12 nA at 1.0 V. Even at 0 V, a photocurrent of 8.5 nA was observed. The low onset potential of InGaN/GaN is in accordance with its low flat band potential [40,41], indicating that the electrode can become depleted by applying a low bias, facilitating the separation of photo-induced charge carriers. Therefore, it provides the possibility for measurements at zero applied bias.

3.2. Photoelectrochemical Imaging Using InGaN

Figure 6a,b shows the photocurrent images of a PMMA dot on the InGaN surface with a modulation frequency of 1 kHz using a focused laser beam at a bias of 0.6 V and 0 V (vs. Ag/AgCl), respectively. The polymer dots were clearly observed in the photocurrent images, with decreased photocurrent values compared to a blank surface area owing to the high impedance of the PMMA dot. The image in Figure 6a shows a significant gradient of the photocurrent across the uncoated area exposed to electrolyte. This can be attributed to the sample not being mounted perfectly perpendicular to the incoming laser beam resulting in a change of the focused laser spot size across the sample. Where applications require imaging over a large area, a tilt module for straightening the sample would have to be integrated into the experimental setup. However, for imaging over small distances, this effect becomes negligible, as will become clear in the next section. The images in Figure 6b and, less obviously, in Figure 6a display a periodic pattern in the photocurrent distribution. It is assumed that this is caused by a striation effect in the InGaN/GaN substrate similar to the one previously observed in silicon [42]. It is worth noting that the ability to image at 0 V will broaden the application of this technique in biological systems, and also possesses an advantage from an energy perspective. To measure the lateral resolution, a photocurrent line scan across the edge of the polymer film was recorded with a focused laser beam and 1 μm step size (Figure 6c). The lateral resolution is derived from the full width at half maximum (FWHM) value of the first derivative of the line [43] (Figure 6d), which is 7 μm for InGaN. This result could be due to a weak adhesion between PMMA and the InGaN surface, thus not giving a steep edge of the polymer, or light scattering within the structure. The diffusion length of minority charge carriers in InGaN should not affect the resolution, as it is less than 200 nm and decreases with increasing indium content [44]. Hence, InGaN is promising for the production of photocurrent images with a higher resolution.

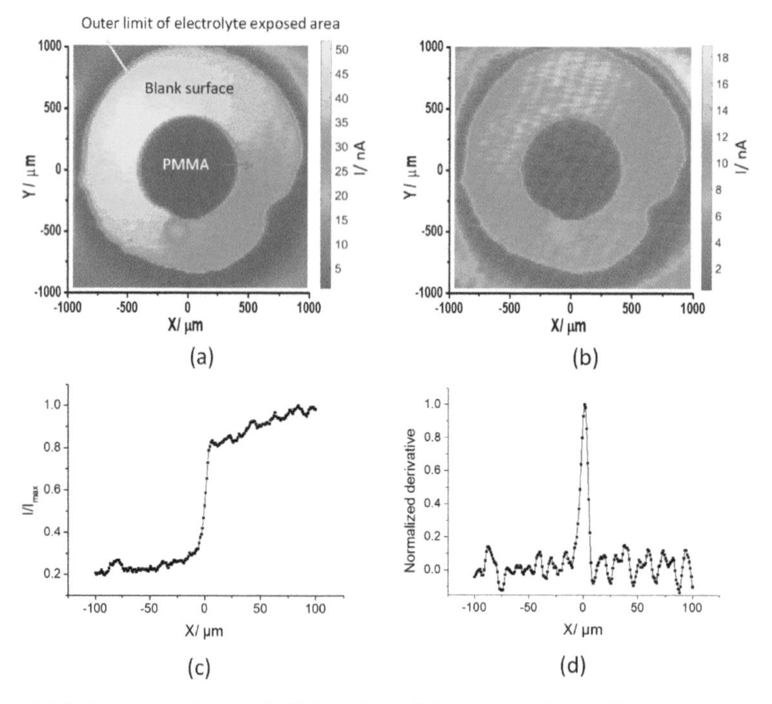

Figure 6. AC photocurrent images of a PMMA dot on InGaN measured at 0.6 V (**a**) and 0 V (**b**); X axis line scan across the polymer edge (indicated by the red arrow in (**a**)) at 0.6 V (**c**) and its corresponding first derivative plot (**d**).

3.3. Cell Imaging on InGaN

Figure 7a shows a photocurrent image of an MG-63 cell seeded on the InGaN surface obtained at a bias of 1.05 V, with a light modulation frequency of 1 kHz. The cell profile is clearly observed, as the photocurrent is smaller in the cell attachment area than on the blank surface. Both the photocurrent image and the corresponding optical image (Figure 7b) show good correlation. Apart from the cell in the center of the image (outline superimposed in blue in Figure 7a), another three cells are visible towards the edges (outlines superimposed in red in Figure 7a). As the latter cells are rounded, it can be assumed that they are not attached to the sensor surface and do therefore not cause a significant change in the local photocurrent. The photocurrent under a cell attached to the semiconductor surface is affected by the narrow gap (> 10 nm) formed between the cell membrane and the surface, as described previously for cells cultured on ITO [24]. The photocurrent is caused by the oxidation of hydroxide. Transport of hydroxide to the surface is hindered by diffusion into the narrow electrolyte gap between cell and surface, thereby reducing the photocurrent under the cell. The negative surface charge of the cell causes an additional reduction in the transport of hydroxide ions to the surface. Hence, a correlation between the photocurrent and the cell surface charge was found [24].

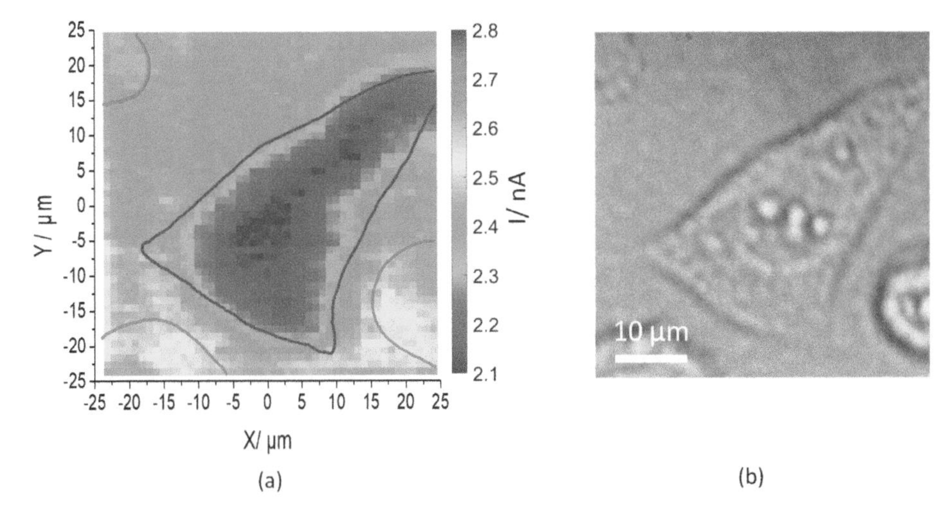

Figure 7. (**a**) AC photocurrent image of a mesenchymal stem cell on InGaN surface (cell shapes from (**b**) superimposed in blue for an attached cell and red for non-attached cells); and (**b**) its corresponding optical image.

3.4. Cell Viability

To check the invasiveness of InGaN-based AC photocurrent imaging, cell viability for cells with and without AC photocurrent raster scan were tested (Figure 8). Calcein AM can permeate through intact cell membranes and react with the intracellular enzyme esterase, giving an intensely green fluorescence in live cells (excitation/emission 495 nm/515 nm). Ethidium homodimer-1 only passes through disrupted membranes, emitting intense red fluorescence in dead cells upon binding to nucleic acids (excitation/emission 495 nm/635 nm). Hoechst stain is a cell-permeant nuclear counterstain that emits blue fluorescence when bound to dsDNA (excitation/emission 350 nm/461 nm) to determine cell numbers. Results show that 98.92% ± 0.15% MG-63 cells on the surface were viable after a photocurrent raster scan compared to 98.97% ± 0.11% on a control sample, indicating that this imaging technique has no negative effect on the cells.

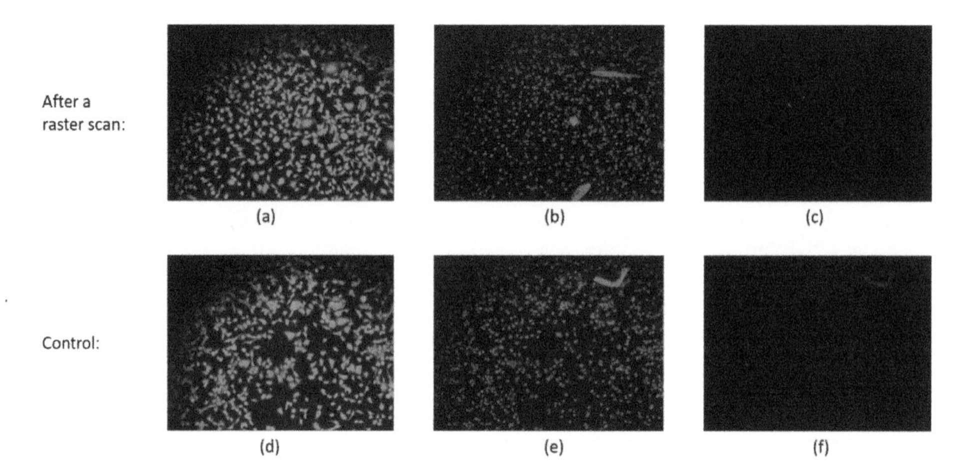

Figure 8. (**a**,**b**,**c**) Fluorescence microscope images of MG-63 cells taken after photocurrent imaging, living cells with intact membranes appeared green, dead cells with collapsed membrane appeared red, and the nuclei of the cells appeared blue. (**d**,**e**,**f**) Images of MG-63 cells that were not subjected to imaging.

4. Conclusions

An $In_{0.175}Ga_{0.825}N$/GaN structure on sapphire was investigated as a substrate for photocurrent imaging without any modification. It showed a considerable photocurrent under illumination with a 405 nm diode laser. Clear photocurrent images of a PMMA dot were obtained with a focused laser beam at 1 kHz modulation frequency, indicating a unique advantage over ITO and ZnO studied previously. In addition, photocurrent imaging at a low bias (0 V) was demonstrated and photocurrent imaging of a cell was achieved, showing a great potential of InGaN for applications in bioimaging and biosensing.

Author Contributions: B.Z. carried out the measurements and prepared the original draft. A.D. contributed to the data analysis and edited the paper. M.J.K. fabricated the InGaN/GaN, carried out photoluminescence and resistance measurements and co-wrote the paper. R.A.O., C.J.H. and S.K. contributed to the experimental design and the interpretation of data and co-wrote the paper.

Funding: This research was funded by the China Scholarship Council (PhD studentship to B. Zhou), Horizon 2020 (a Marie Skłodowska-Curie Individual Fellowship to A. Das, H2020-MSCA-IF-2016-745820) and the Engineering and Physical Science Research Council (EPSRC, EP/R035571/1).

Conflicts of Interest: The authors declare no conflict of interest.

References

1. Hafeman, D.G.; Parce, J.W.; McConnell, H.M. Light-addressable potentiometric sensor for biochemical systems. *Science* **1988**, *240*, 1182–1185. [CrossRef] [PubMed]
2. Mourzina, Y.; Yoshinobu, T.; Schubert, J.; Lüth, H.; Iwasaki, H.; Schöning, M.J. Ion-selective light-addressable potentiometric sensor (LAPS) with chalcogenide thin film prepared by pulsed laser deposition. *Sens. Actuators B Chem.* **2001**, *80*, 136–140. [CrossRef]
3. Oba, N.; Yoshinobu, T.; Iwasaki, H. Redox Potential Imaging Sensor. *Jpn. J. Appl. Phys.* **1996**, *35*, 460–463. [CrossRef]
4. Seki, A.; Ikeda, S.; Kubo, I.; Karube, I. Biosensors based on light-addressable potentiometric sensors for urea, penicillin and glucose. *Anal. Chim. Acta* **1998**, *373*, 9–13. [CrossRef]
5. Poghossian, A.; Ingebrandt, S.; Offenhäusser, A.; Schöning, M.J. Field-effect devices for detecting cellular signals. *Semin. Cell Dev. Biol.* **2009**, *20*, 41–48. [CrossRef] [PubMed]

6. Hu, N.; Wu, C.; Ha, D.; Wang, T.; Liu, Q.; Wang, P. A novel microphysiometer based on high sensitivity LAPS and microfluidic system for cellular metabolism study and rapid drug screening. *Biosens. Bioelectron.* **2013**, *40*, 167–173. [CrossRef] [PubMed]
7. Dantism, S.; Takenaga, S.; Wagner, T.; Wagner, P.; Schöning, M.J. Differential imaging of the metabolism of bacteria and eukaryotic cells based on light-addressable potentiometric sensors. *Electrochim. Acta* **2017**, *246*, 234–241. [CrossRef]
8. Owicki, J.C.; Bousse, L.J.; Hafeman, D.G.; Kirk, G.L.; Olson, J.D.; Wada, H.G.; Parce, J.W. The Light-Addressable Potentiometric Sensor: Principles and Biological Applications. *Annu. Rev. Biophys. Biomol. Struct.* **1994**, *23*, 87–114. [CrossRef]
9. Wu, F.; Campos, I.; Zhang, D.W.; Krause, S. Biological imaging using light-addressable potentiometric sensors and scanning photo-induced impedance microscopy. *Proc. R. Soc. A Math. Phys. Eng. Sci.* **2017**, *473*, 2201. [CrossRef]
10. Ito, Y. High-spatial resolution LAPS. *Sens. Actuators B-Chem.* **1998**, *52*, 107–111. [CrossRef]
11. Krause, S.; Talabani, H.; Xu, M.; Moritz, W.; Griffiths, J. Scanning photo-induced impedance microscopy—An impedance based imaging technique. *Electrochim. Acta* **2002**, *47*, 2143–2148. [CrossRef]
12. Chen, L.; Zhou, Y.; Jiang, S.; Kunze, J.; Schmuki, P.; Krause, S. High resolution LAPS and SPIM. *Electrochem. commun.* **2010**, *12*, 758–760. [CrossRef]
13. Moritz, W.; Gerhardt, I.; Roden, D.; Xu, M.; Krause, S. Photocurrent measurements for laterally resolved interface characterization. *Fresenius J. Anal. Chem.* **2000**, *367*, 329–333. [CrossRef] [PubMed]
14. Das, A.; Chang, L.B.; Lai, C.S.; Lin, R.M.; Chu, F.C.; Lin, Y.H.; Chow, L.; Jeng, M.J. GaN Thin Film Based Light Addressable Potentiometric Sensor for pH Sensing Application. *Appl. Phys. Express* **2013**, *6*, 3. [CrossRef]
15. Suzurikawa, J.; Nakao, M.; Jimbo, Y.; Kanzaki, R.; Takahashi, H. A light addressable electrode with a TiO_2 nanocrystalline film for localized electrical stimulation of cultured neurons. *Sens. Actuators, B Chem.* **2014**, *192*, 393–398. [CrossRef]
16. Yang, C.; Chen, C.; Chang, L.; Lai, C. IGZO Thin-Film Light-Addressable Potentiometric Sensor. *IEEE Electron Device Lett.* **2016**, *37*, 1481–1484. [CrossRef]
17. Wang, J.; Wu, F.; Watkinson, M.; Zhu, J.; Krause, S. "click" Patterning of Self-Assembled Monolayers on Hydrogen-Terminated Silicon Surfaces and Their Characterization Using Light-Addressable Potentiometric Sensors. *Langmuir* **2015**, *31*, 9646–9654. [CrossRef]
18. Wang, J.; Zhou, Y.; Watkinson, M.; Gautrot, J.; Krause, S. High-sensitivity light-addressable potentiometric sensors using silicon on sapphire functionalized with self-assembled organic monolayers. *Sens. Actuators B Chem.* **2015**, *209*, 230–236. [CrossRef]
19. Wu, F.; Zhang, D.W.; Wang, J.; Watkinson, M.; Krause, S. Copper Contamination of Self-Assembled Organic Monolayer Modified Silicon Surfaces Following a "Click" Reaction Characterized with LAPS and SPIM. *Langmuir* **2017**, *33*, 3170–3177. [CrossRef]
20. Wang, J.; Campos, I.; Wu, F.; Zhu, J.; Sukhorukov, G.B.; Palma, M.; Watkinson, M.; Krause, S. The effect of gold nanoparticles on the impedance of microcapsules visualized by scanning photo-induced impedance microscopy. *Electrochim. Acta* **2016**, *208*, 39–46. [CrossRef]
21. Zhang, D.W.; Wu, F.; Wang, J.; Watkinson, M.; Krause, S. Image detection of yeast Saccharomyces cerevisiae by light-addressable potentiometric sensors (LAPS). *Electrochem. Commun.* **2016**, *72*, 41–45. [CrossRef]
22. Vogel, Y.B.; Gooding, J.J.; Ciampi, S. Light-addressable electrochemistry at semiconductor electrodes: Redox imaging, mask-free lithography and spatially resolved chemical and biological sensing. *Chem. Soc. Rev.* **2019**, 3723–3739. [CrossRef] [PubMed]
23. Zhang, D.W.; Wu, F.; Krause, S. LAPS and SPIM Imaging Using ITO-Coated Glass as the Substrate Material. *Anal. Chem.* **2017**, *89*, 8129–8133. [CrossRef] [PubMed]
24. Wu, F.; Zhou, B.; Wang, J.; Zhong, M.; Das, A.; Watkinson, M.; Hing, K.; Zhang, D.W.; Krause, S. Photoelectrochemical Imaging System for the Mapping of Cell Surface Charges. *Anal. Chem.* **2019**, *91*, 5896–5903. [CrossRef] [PubMed]
25. Tu, Y.; Ahmad, N.; Briscoe, J.; Zhang, D.W.; Krause, S. Light-Addressable Potentiometric Sensors Using ZnO Nanorods as the Sensor Substrate for Bioanalytical Applications. *Anal. Chem.* **2018**, *90*, 8708–8715. [CrossRef] [PubMed]
26. Nakamura, S.; Senoh, M.; Mukai, T. P-GaN/N-InGaN/N-GaN Double-Heterostructure Blue-Light-Emitting Diodes. *Jpn. J. Appl. Phys.* **1993**, *32*, 8–11. [CrossRef]

27. Lin, H.W.; Lu, Y.J.; Chen, H.Y.; Lee, H.M.; Gwo, S. InGaN/GaN nanorod array white light-emitting diode. *Appl. Phys. Lett.* **2010**, *97*, 98–101.
28. Matioli, E.; Neufeld, C.; Iza, M.; Cruz, S.C.; Al-Heji, A.A.; Chen, X.; Farrell, R.M.; Keller, S.; DenBaars, S.; Mishra, U.; et al. High internal and external quantum efficiency InGaN/GaN solar cells. *Appl. Phys. Lett.* **2011**, *98*, 2009–2012. [CrossRef]
29. Kibria, M.G.; Chowdhury, F.A.; Zhao, S.; AlOtaibi, B.; Trudeau, M.L.; Guo, H.; Mi, Z. Visible light-driven efficient overall water splitting using p-type metal-nitride nanowire arrays. *Nat. Commun.* **2015**, *6*, 1–8. [CrossRef]
30. Wallys, J.; Teubert, J.; Furtmayr, F.; Hofmann, D.M.; Eickhoff, M. Bias-enhanced optical ph response of group III-nitride nanowires. *Nano Lett.* **2012**, *12*, 6180–6186. [CrossRef]
31. Maier, K.; Helwig, A.; Müller, G.; Becker, P.; Hille, P.; Schörmann, J.; Teubert, J.; Eickhoff, M. Detection of oxidising gases using an optochemical sensor system based on GaN/InGaN nanowires. *Sens. Actuators B Chem.* **2014**, *197*, 87–94. [CrossRef]
32. Riedel, M.; Hölzel, S.; Hille, P.; Schörmann, J.; Eickhoff, M.; Lisdat, F. InGaN/GaN nanowires as a new platform for photoelectrochemical sensors—Detection of NADH. *Biosens. Bioelectron.* **2017**, *94*, 298–304. [CrossRef] [PubMed]
33. Taylor, E.; Fang, F.; Oehler, F.; Edwards, P.R.; Kappers, M.J.; Lorenz, K.; Alves, E.; McAleese, C.; Humphreys, C.J.; Martin, R.W. Composition and luminescence studies of InGaN epilayers grown at different hydrogen flow rates. *Semicond. Sci. Technol.* **2013**, *28*, 6. [CrossRef]
34. Martin, R.W.; Edwards, P.R.; O'Donnell, K.P.; Mackay, E.G.; Watson, I.M. Microcomposition and Luminescence of InGaN Emitters. *Phys. Status Solidi* **2002**, *192*, 117–123. [CrossRef]
35. Tauc, J.; Grigorovici, R.; Vancu, A. Optical Properties and Electronic Structure of Amorphous Germanium. *Phys. Status Solidi* **1966**, *15*, 627–637. [CrossRef]
36. Viezbicke, B.D.; Patel, S.; Davis, B.E.; Birnie III, D.P. Evaluation of the Tauc method for optical absorption edge determination: ZnO thin films as a model system. *Phys. Status Solidi* **2015**, *252*, 1700–1710. [CrossRef]
37. Brown, G.F.; Iii, J.W.A.; Walukiewicz, W.; Wu, J. Solar Energy Materials & Solar Cells Finite element simulations of compositionally graded InGaN solar cells. *Sol. Energy Mater. Sol. Cells* **2010**, *94*, 478–483.
38. Can, M.; Havare, A.K.; Aydın, H.; Yagmurcukardes, N.; Demic, S.; Icli, S.; Okur, S. Electrical properties of SAM-modified ITO surface using aromatic small molecules with double bond carboxylic acid groups for OLED applications. *Appl. Surf. Sci.* **2014**, *314*, 1082–1086. [CrossRef]
39. Hammer, M.S.; Deibel, C.; Pflaum, J.; Dyakonov, V. Effect of doping of zinc oxide on the hole mobility of poly (3-hexylthiophene) in hybrid transistors. *Org. Electron.* **2010**, *11*, 1569–1577. [CrossRef]
40. Ebaid, M.; Kang, J.H.; Lim, S.H.; Ha, J.S.; Lee, J.K.; Cho, Y.H.; Ryu, S.W. Enhanced solar hydrogen generation of high density, high aspect ratio, coaxial InGaN/GaN multi-quantum well nanowires. *Nano Energy* **2015**, *12*, 215–223. [CrossRef]
41. Kobayashi, N.; Morita, R.; Narumi, T.; Yamamoto, J.; Ban, Y.; Wakao, K. Flat-band potentials of GaN and InGaN/GaN QWs by bias-dependent photoluminescence in electrolyte solution. *J. Cryst. Growth* **2007**, *298*, 515–517. [CrossRef]
42. Miyamoto, K.; Sugawara, Y.; Kanoh, S.; Yoshinobu, T.; Wagner, T.; Schöning, M.J. Image correction method for the chemical imaging sensor. *Sens. Actuators B Chem.* **2010**, *144*, 344–348. [CrossRef]
43. George, M.; Parak, W.J.; Gerhardt, I.; Moritz, W.; Kaesen, F.; Geiger, H.; Eisele, I.; Gaub, H.E. Investigation of the spatial resolution of the light-addressable potentiometric sensor. *Sens. Actuators A Phys.* **2000**, *86*, 187–196. [CrossRef]
44. Kumakura, K.; Makimoto, T.; Kobayashi, N.; Hashizume, T.; Fukui, T.; Hasegawa, H. Minority carrier diffusion lengths in MOVPE-grown n- and p-InGaN and performance of AlGaN/InGaN/GaN double heterojunction bipolar transistors. *J. Cryst. Growth* **2007**, *298*, 787–790. [CrossRef]

Article

Modeling of the Return Current in a Light-Addressable Potentiometric Sensor

Tatsuo Yoshinobu [1,2,*], Daisuke Sato [2], Yuanyuan Guo [3], Carl Frederik Werner [2] and Ko-ichiro Miyamoto [2]

[1] Department of Biomedical Engineering, Tohoku University, 6-6, Aza-Aoba, Aramaki, Aoba-ku, Sendai 980-8579, Japan
[2] Department of Electronic Engineering, Tohoku University, 6-6, Aza-Aoba, Aramaki, Aoba-ku, Sendai 980-8579, Japan; d.sato@ecei.tohoku.ac.jp (D.S.); werner@ecei.tohoku.ac.jp (C.F.W.); k-miya@ecei.tohoku.ac.jp (K.-i.M.)
[3] Frontier Research Institute of Interdisciplinary Sciences, Tohoku University, 6-3 Aza-Aoba, Aramaki, Aoba-ku, Sendai 980-8578, Japan; yyuanguo@fris.tohoku.ac.jp
* Correspondence: nov@ecei.tohoku.ac.jp

Received: 24 September 2019; Accepted: 18 October 2019; Published: 21 October 2019

Abstract: A light-addressable potentiometric sensor (LAPS) is a chemical sensor with a field-effect structure based on semiconductor. Its response to the analyte concentration is read out in the form of a photocurrent generated by illuminating the semiconductor with a modulated light beam. As stated in its name, a LAPS is capable of spatially resolved measurement using a scanning light beam. Recently, it has been pointed out that a part of the signal current is lost by the return current due to capacitive coupling between the solution and the semiconductor, which may seriously affect the sensor performance such as the signal-to-noise ratio, the spatial resolution, and the sensitivity. In this study, a circuit model for the return current is proposed to study its dependence on various parameters such as the diameter of contact area, the modulation frequency, the specific conductivity of the solution, and the series resistance of the circuit. It is suggested that minimization of the series resistance of the circuit is of utmost importance in order to avoid the influence of the return current. The results of calculation based on this model are compared with experimental results, and its applicability and limitation are discussed.

Keywords: light-addressable potentiometric sensor; LAPS; chemical imaging sensor; field-effect sensor

---

1. Introduction

A light-addressable potentiometric sensor (LAPS) [1–3] is a semiconductor-based chemical sensor, which has a field-effect structure shown in Figure 1a. A dc voltage is applied to induce a depletion layer, the thickness of which varies due to the field effect in response to the analyte concentration on the sensing surface. A photocurrent generated by illuminating the semiconductor is measured to detect the variation of the capacitance of the depletion layer and to determine the analyte concentration. A spatially resolved measurement is possible by using a scanning light beam, which is the basis of the chemical imaging sensor [3,4] and the scanning photo-induced impedance microscopy [5].

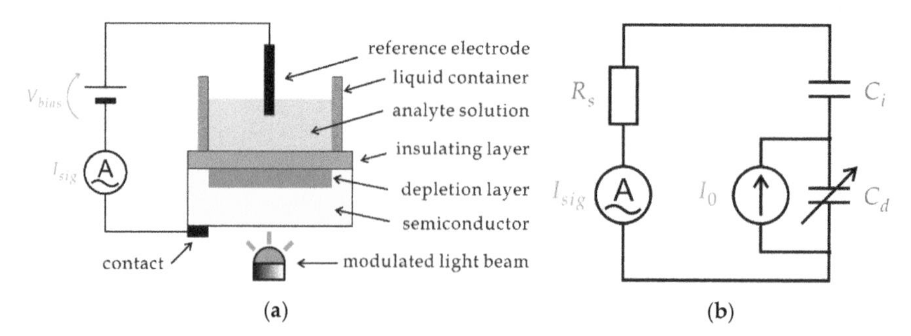

Figure 1. (**a**) Schematic of a LAPS. A bias voltage is applied to the field-effect structure so that a depletion layer is formed. The thickness of the depletion layer varies with the analyte concentration. The semiconductor substrate is illuminated with a modulated light beam, and the ac photocurrent signal I_{sig} is measured and correlated to the analyte concentration. (**b**) A circuit model of a LAPS. The internal current I_0 is divided by the capacitance of the depletion layer C_d and that of the insulating layer C_i connected to the input resistance of the circuit R_s.

In theoretical analysis of a LAPS, a simple circuit model, shown in Figure 1b, has been conventionally employed [6–8]. In this model, separation of electrons and holes by the electric field inside the depletion layer is represented by an internal ac current source I_0, which is divided by the capacitance of the depletion layer C_d and that of the insulating layer C_i connected to the series resistance of the circuit R_s. Here, R_s consists of the resistance of the solution between the illuminated point and the reference electrode, the resistance of the reference electrode (in case the counter electrode is not used), the input resistance of the ammeter, and the contact resistance on the back surface of the semiconductor substrate. When the thickness of the depletion layer changes in response to the analyte concentration of the solution in contact with the sensing surface, variation of C_d results in variation of the signal current I_{sig}. This circuit model was further combined with the carrier diffusion model [8–13] to describe the operation of the chemical imaging sensor, taking account of lateral diffusion and recombination of minority carriers inside the semiconductor substrate.

Although these models were successful in describing important features of the chemical imaging sensor including its spatial resolution and frequency characteristics, Poghossian et al. [14] pointed out the influence of capacitive coupling between the solution and the semiconductor substrate in the non-illuminated region, which was not included in existing models. Figure 2 shows the simplest model, in which a part of the ac current returns to the semiconductor substrate through C_i' and C_d' without contributing to the signal current I_{sig}. It should be noted that this effect applies only to an ac current. In light-activated electrochemistry (LAE) [15], which has a similar setup to that of LAPS but uses dc faradaic current, the high impedance of the non-illuminated region separates the solution and the substrate. Figure 2 gives only an intuition that the capacitive coupling increases with the area of the non-illuminated region and the frequency of the ac current. It can be easily speculated that a loss of the signal current due to the return current may have a large impact on the signal-to-noise ratio, the spatial resolution, and the sensitivity. To be able to understand the dependence of the return current on various parameters and to evaluate its impact on the sensor performance, the model shown in Figure 2 is far too simple.

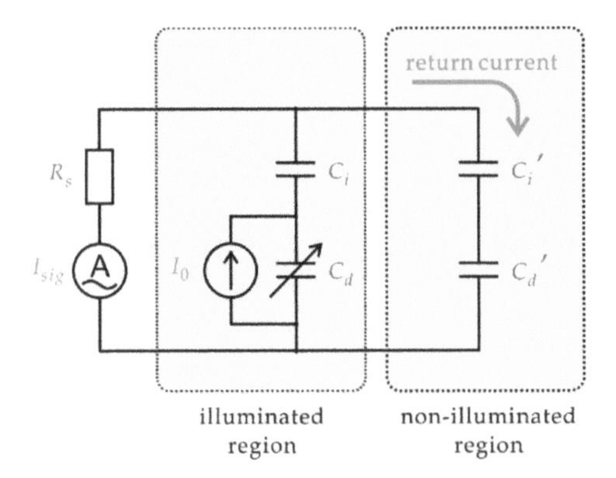

Figure 2. A simple circuit model of the return current. Due to the capacitive coupling of the solution and the semiconductor substrate, a part of the ac photocurrent returns to the semiconductor substrate through the capacitance of the insulating layer and that of the depletion layer in the non-illuminated region without contributing to the signal current I_{sig}.

In this study, a new circuit model is proposed, in which the path of the return current is described as a transmission line. The model is used to calculate the influence of the return current on I_{sig} and its dependence on parameters such as the frequency, the size of the non-illuminated region, the conductivity of the solution, and the series resistance of the circuit. Applicability and limitation of the model are discussed by comparing the results obtained by calculation and measurement.

2. Model

Figure 3a shows the top view of the model, in which a circular region on the sensing surface with a radius R is in contact with the solution and a circular region with a radius r_0 at the center is illuminated. The rest of the contact area ($r_0 < r < R$) is non-illuminated. The resistance of the solution in an infinitesimal volume between the inner and outer walls of a hollow cylinder with a radius r and a thickness dr shown in the upper part of Figure 3b is given by:

$$\frac{dr}{2\pi rh\sigma}, \tag{1}$$

where h is the height of the solution and σ is the specific conductivity of the solution. The combined capacitance of the insulating layer and the depletion layer in an infinitesimal area between r and $r + dr$ shown in the lower part of Figure 3b is given by

$$2\pi rcdr, \tag{2}$$

where c is the combined capacitance per unit area. When the capacitance of the insulating layer per unit area c_i and that of the depletion layer c_d are connected in series, the combined capacitance is given by

$$c = \frac{1}{\frac{1}{c_i} + \frac{1}{c_d}}. \tag{3}$$

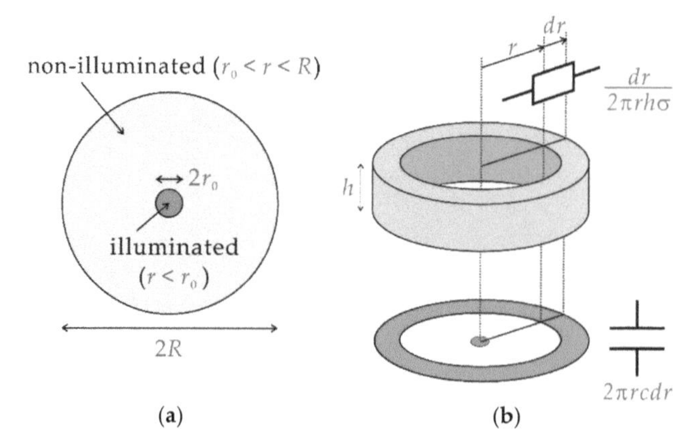

Figure 3. (**a**) Top view of the sensing surface in contact with the solution. A part of the photocurrent generated inside the illuminated region ($r < r_0$) returns to the semiconductor substrate through the non-illuminated region ($r_0 < r < R$) by capacitive coupling. (**b**) The resistance of the solution between the inner and outer walls of a hollow cylinder and the combined capacitance of the insulating layer and the depletion layer under the sensing surface in a ring shape are considered.

The admittance of the non-illuminated region Y is represented as a ladder network shown in Figure 4a, which is essentially a finite-length transmission line with an open end. Unlike a conventional transmission line, however, the resistance and the capacitance per unit length are not constant but dependent on r as described by (1) and (2), respectively. The telegraph equations of the transmission line are

$$I'(r) = -j\omega 2\pi rcV(r), \tag{4a}$$

$$V'(r) = -\frac{1}{2\pi rh\sigma}I(r), \tag{4b}$$

where $I(r)$ and $V(r)$ are complex numbers representing the phasors of the ac current and the ac voltage with an angular frequency ω $(= 2\pi f)$ at position r and j is the imaginary unit. From Equations (4a) and (4b) we obtain a second-order ordinary differential equation,

$$I''(r) = \frac{1}{r}I'(r) + jkI(r), \tag{5}$$

where

$$k = \frac{\omega c}{h\sigma}. \tag{6}$$

By solving Equation (5) under initial conditions at $r = R$, we can obtain final values $I(r_0)$ and $I'(r_0)$. Then, we also obtain $V(r_0)$ from Equation (4a), and the input admittance Y in Figure 4b is given by

$$Y = \frac{I(r_0)}{V(r_0)}. \tag{7}$$

Once Y is obtained, the signal current I_{sig} can be calculated as follows.

$$I_{sig} = I_0 \times \frac{1}{\left(1 + \frac{C_{d0}}{C_{i0}}\right)(1 + R_sY) + j\omega C_{d0}R_s}. \tag{8}$$

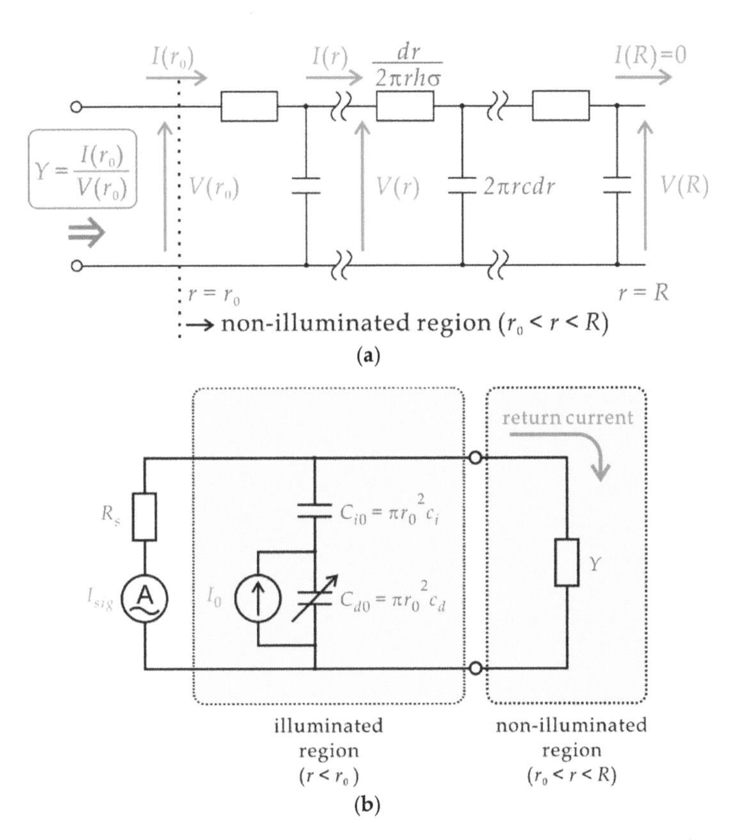

Figure 4. (**a**) A circuit model of the non-illuminated region. The resistance of the solution in an infinitesimal volume and the capacitance in an infinitesimal area shown in Figure 3b are connected in a ladder network in the range of $r_0 < r < R$. By solving the telegraph equations of the transmission line under the initial conditions at $r = R$, the final values at $r = r_0$ are obtained, from which the input admittance Y is obtained. (**b**) A circuit model, in which the path of the return current is represented by admittance Y.

Here, C_{i0} is the capacitance of the insulating layer inside the illuminated region and C_{d0} is that of the depletion layer, which are given by

$$C_{i0} = \pi {r_0}^2 c_i, \tag{9}$$

$$C_{d0} = \pi {r_0}^2 c_d. \tag{10}$$

The initial conditions at $r = R$ are given as follows. Since the current does not flow out of the contact area,

$$I(R) = 0. \tag{11}$$

The value of $I'(R)$, or equivalently the value of $V(R)$, can be arbitrarily given, as we are interested only in the ratio of $I(r)$ and $V(r)$. For simplicity, we choose it to be

$$I'(R) = -1. \tag{12}$$

For ease of calculation, the second-order ordinary differential equation of a complex-valued function (5) can be converted into a set of first-order ordinary differential equations of four real-valued functions by defining

$$y_1(r) = \mathrm{Re}\, I(r), \tag{13a}$$

$$y_2(r) = \mathrm{Im}\, I(r), \tag{13b}$$

$$y_3(r) = y_1{}'(r), \tag{13c}$$

$$y_4(r) = y_2{}'(r). \tag{13d}$$

Then, our problem is to solve a set of differential equations

$$y_1{}'(r) = y_3(r), \tag{14a}$$

$$y_2{}'(r) = y_4(r), \tag{14b}$$

$$y_3{}'(r) = \frac{1}{r} y_3(r) - k y_2(r), \tag{14c}$$

$$y_4{}'(r) = \frac{1}{r} y_4(r) + k y_1(r), \tag{14d}$$

under the initial conditions

$$y_1(R) = 0, \tag{15a}$$

$$y_2(R) = 0, \tag{15b}$$

$$y_3(R) = -1, \tag{15c}$$

$$y_4(R) = 0. \tag{15d}$$

Finally, we obtain the values $y_1(r_0)$, $y_2(r_0)$, $y_3(r_0)$, and $y_4(r_0)$, which give

$$I(r_0) = y_1(r_0) + j y_2(r_0), \tag{16a}$$

$$V(r_0) = \frac{1}{\omega 2\pi r_0 c}\{-y_4(r_0) + j y_3(r_0)\}. \tag{16b}$$

In the following sections, calculations were done by a Runge–Kutta solver *ode45* of MATLAB® (MathWorks). Parameters listed in Table 1 were used so that the results of calculation can be compared with those experimentally obtained.

Table 1. Parameters used in calculation unless otherwise specified.

Parameter	Symbol	Value	Unit
Capacitance of the insulating layer per unit area	c_i	4.49×10^{-4}	F/m^2
Capacitance of the depletion layer per unit area	c_d	2.42×10^{-4}	F/m^2
Specific conductivity of the solution	σ	2	mS/cm
Series resistance of the circuit	R_s	1800	Ω
Radius of the illuminated region	r_0	0.5	mm

3. Dependence on R and f

First of all, the dependence of the admittance Y on the radius of the non-illuminated region R and the frequency f was calculated as summarized in Figure 5.

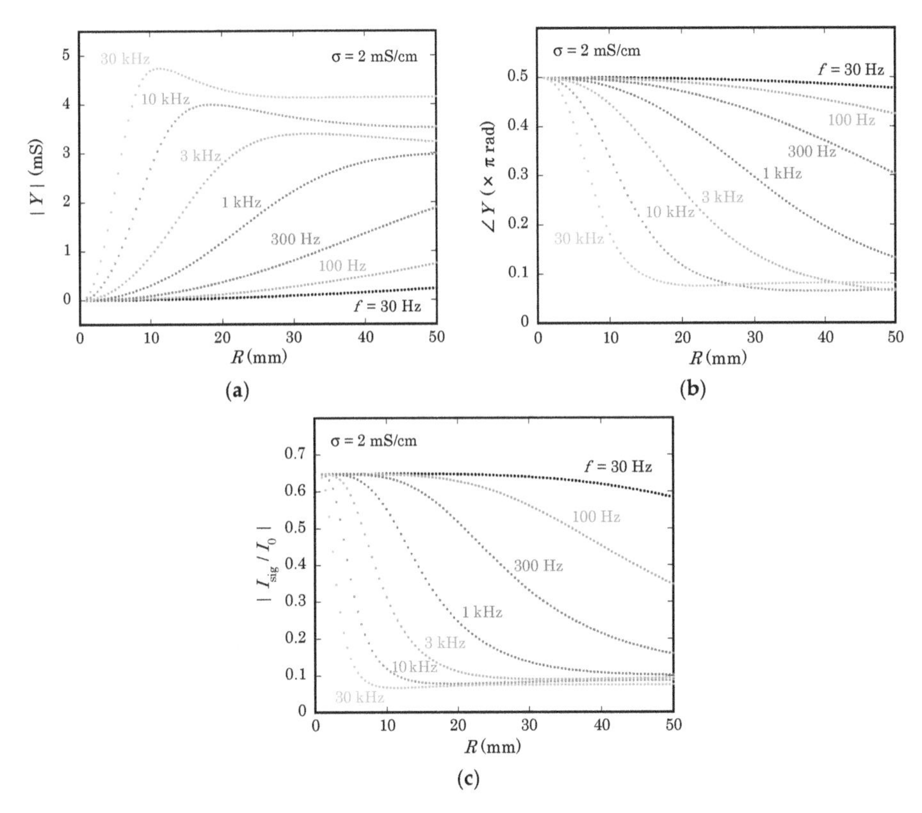

Figure 5. Dependence of (**a**) the magnitude and (**b**) the argument of the input admittance of the non-illuminated region Y, and (**c**) the ratio of the signal current I_{sig} to the internal current I_0 on the radius of the non-illuminated region R at different modulation frequencies of the light beam f.

Figure 5a shows the magnitude of Y calculated with the model described in the previous section. As expected, the magnitude of Y becomes larger at higher frequencies, meaning that more current returns from the solution to the semiconductor substrate through their capacitive coupling. Moreover, the magnitude of Y becomes larger as R increases and the contact area becomes larger. However, the admittance of a finite-length transmission line does not always increase monotonously with its length. In fact, curves for the frequency of 3 kHz and higher have maxima.

In Figure 5b, the argument of Y is plotted as a function of R. When the contact area is small, the transmission line is mostly capacitive, and it becomes more resistive as R increases. This behavior is qualitatively understood as follows. When R increases, more portion of the return current flows through the capacitance at locations further from the center, in other words, after going through a larger lateral resistance of the solution on the way. At higher frequencies, this transition of the argument of Y occurs at smaller distance R.

Figure 5c shows the magnitude of I_{sig}/I_0 as a function of R at different frequencies. As R increases, the admittance of the non-illuminated region Y becomes larger and the signal current decreases due to the increase of the return current. In the present case, where calculation was done with parameters listed in Table 1, the signal current decreased down to about 14% of the value at $R = r_0$, where the non-illuminated region does not exist.

4. Dependence on σ and R_S

In Figure 6a, the dependence of the signal current I_{sig} on the radius of the non-illuminated region R is plotted for different values of the specific conductivity of the solution σ. Here, the series resistance of the circuit R_S and the frequency f were kept constant at 1800 Ω and 1 kHz, respectively. When the solution is less conductive, the effect of the return current is relatively smaller, because the lateral resistance will limit the distance from the center, within which the capacitive coupling contributes to the return current. It should be noted, however, that a smaller conductivity of the solution also implies a higher resistance of the solution between the illuminated point and the reference electrode, which increases R_S. As a whole, therefore, less conductivity will not necessarily reduce the effect of the return current.

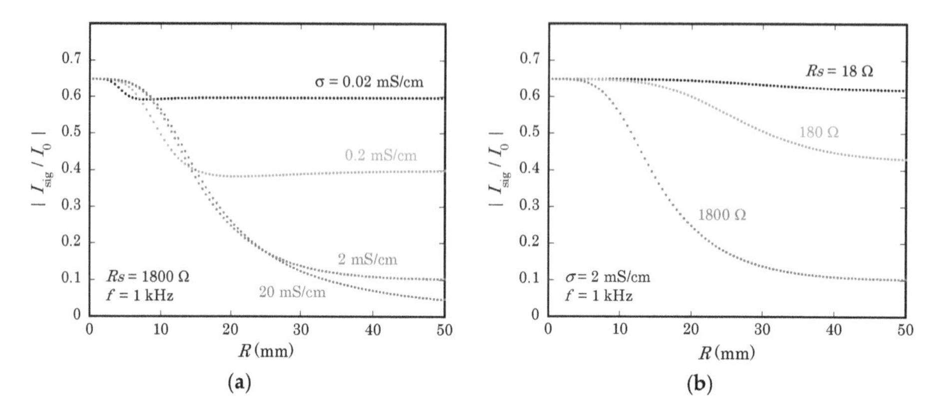

Figure 6. The magnitude of I_{sig} calculated for different values of (**a**) the specific conductivity of the solution σ and (**b**) the resistance of the circuit R_S.

In Figure 6b, calculation was done for different values of R_S, while keeping σ and f at 2 mS/cm and 1 kHz, respectively. It clearly shows that reduction of R_S is of utmost importance in removing the effect of the return current. To reduce R_S, a three-electrode system with a counter electrode should be used to bypass the resistance of the reference electrode. The electronic circuit to collect the ac photocurrent signal, typically a transimpedance amplifier based on operational amplifiers, should be carefully designed to minimize its input impedance. The ohmic contact on the back surface of the semiconductor substrate must be carefully formed to minimize the contact resistance.

In some applications, a planer counter electrode can be placed in parallel to the sensing surface, so that the vertical distance from the illuminated point to the counter electrode is always small and constant even when the position of the light beam is moved for scanning. In case of measurement inside a microfluidic device, a metallic wire can be inserted along the microchannel as a counter electrode. In these cases, the counter electrode helps, on one hand, to reduce R_S, but it also shortcuts the lateral resistance of the solution and delivers the return current to locations far from the illuminated point and may increase the return current.

5. Impact on the Sensitivity

In chemical imaging based on a LAPS, a focused light beam scans the semiconductor substrate, and the signal current is recorded at each pixel. The signal current is then converted into the analyte concentration using a calibration curve acquired prior to the measurement. For a small change, a linear approximation is used to convert a variation of the signal current into that of the potential, which is then linearly correlated to the logarithm of the activity of the analyte using the Nernst equation.

Under the existence of a return current, however, this conversion may be systematically affected due to the following reason. During the calibration step, the entire sensing surface is uniformly in contact with known concentrations of analyte solutions. In such a case, the thickness of the depletion layer varies equally both in the illuminated region and in the non-illuminated region, and I_{sig} varies under the global change of c_d. During the measurement step, however, c_d may change only locally, and the return current may be different from that in the calibration step.

To illuminate the difference, calculation was carried out in two different situations. First, I_{sig} was calculated while changing c_d both in the illuminated region and in the non-illuminated region, which corresponds to the situation of the calibration step. Second, I_{sig} was calculated while changing c_d only in the illuminated region with c_d in the non-illuminated region unchanged. Figure 7a shows, for different values of R, the variation of I_{sig} as a function of $\Delta c_d / c_d$ in the range of 0 to 1. It is clearly observed that the variation of I_{sig} for a local change of c_d is smaller than that for a global change of c_d. In other words, a local change of the analyte concentration in imaging will be underestimated due to the difference of the return current during calibration and measurement.

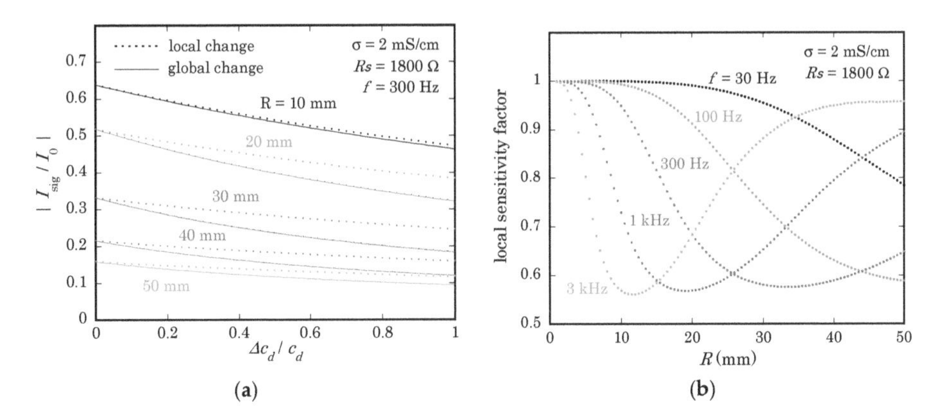

Figure 7. (**a**) Variation of I_{sig} as a function of $\Delta c_d / c_d$ for a global change of c_d (solid lines) and for a local change of c_d (dotted lines). The values of σ, R_s, and f were 2 mS/cm, 1800 Ω, and 300 Hz, respectively. (**b**) The local sensitivity factor calculated for different values of R and f. The values of σ and R_s were 2 mS/cm and 1800 Ω, respectively.

The ratio of the slope (calculated in the range between $\Delta c_d / c_d = 0$ and 1) for a local change of c_d in Figure 7a and that for a global change of c_d was defined as a local sensitivity factor, which indicates the degree of underestimation. Figure 7b shows the local sensitivity factor as a function of R at different frequencies. This result shows that a local change can be underestimated, depending on the combination of R and f, by a factor even smaller than 0.6 in the calculated case, where R_s was 1800 Ω. When R_s was reduced to 18 Ω, the local sensitivity factor calculated within the same range of conditions as in Figure 7b was always larger than 0.96 (data not shown), meaning that the underestimation was less than 4%. This result again shows the importance of reducing R_s in removing the effect of the return current in a LAPS.

6. Comparison with Experiments

A series of experiments were carried out to observe the dependence of I_{sig} on R and f in a real situation. A large-area LAPS plate was prepared by depositing 50 nm SiO_2 and 50 nm Si_3N_4 successively on the entire surface of a 6-inch n-type Si wafer with a thickness of 200 μm and a resistivity of 1–10 Ωcm. An ohmic contact was evaporated on the perimeter of the back surface. To define different sizes of contact areas between the solution and the sensing surface, various sizes of cylindrical liquid containers to accommodate the solution were prepared by a 3D printer and attached to the sensing surface via O-rings with inner diameters 21.0, 40.5, 60.5, 82.5, and 102.5 mm. The solution used in this experiment was 0.1 wt% NaCl solution with a specific conductivity $\sigma = 2.0$ mS/cm and the height was h = 10 mm. A Ag/AgCl reference electrode (RE-1B, BAS Inc.) was dipped into the solution at the center.

The capacitance of the insulating layer per unit area c_i and the series resistance of the circuit R_s were determined by the following method. A coil with an inductance 100 mH was inserted in the circuit and the sensor was biased at V_{bias} = +1.0 V, where the depletion layer disappears. For each size of the contact area, the ac current in response to the application of a small ac voltage was recorded to find the resonance peak while scanning the frequency. The capacitance was calculated from the peak position, and the series resistance was calculated from the peak height. Then, the value of c_i was determined to be 4.49×10^{-4} F/m^2 by linear regression of the capacitance versus the contact area, and the average value of R_s was 1.8×10^3 Ω. The measured value of c_i was close to 4.54×10^{-4} F/m^2, a theoretical value for a double layer comprising 50 nm SiO_2 and 50 nm Si_3N_4. The capacitance of the depletion layer per unit area c_d at V_{bias} = −2.0 V was determined to be 2.42×10^{-4} F/m^2 by linear regression of the measured capacitance versus the contact area. The measured value of c_d was close to 2.28×10^{-4} F/m^2, a theoretical value for n-type Si with a donor concentration $N_D = 4 \times 10^{15}$ cm^{-3} under strong inversion at 300 K.

A LAPS signal was collected at V_{bias} = −2.0 V by illuminating the center of the back surface with a modulated light beam from a red LED. The radius of illumination was restricted to 0.5 mm by an aperture. Figure 8 shows the magnitude of the ac current signal for different values of R and f. Here, it should be noted that the vertical axis of Figure 8 is not normalized, while the magnitude of I_{sig} in Figure 5c is normalized to I_0, which depends on the frequency but cannot be directly measured in experiments. The similar dependence of curves on R and f in Figures 5c and 8 proves that the loss of the current signal is qualitatively reproduced by calculating the return current using the proposed model.

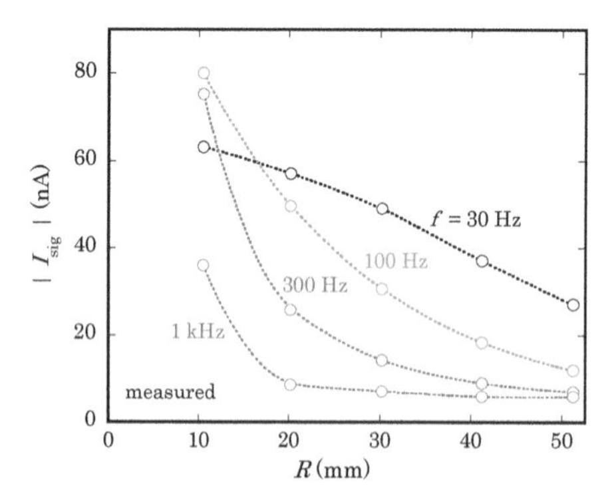

Figure 8. Experimentally obtained values of the magnitude of the ac current signal I_{sig} at different values of the radius of the contact area R and the frequency f. The measurement was done in a 0.1 wt% NaCl solution with a specific conductivity of 2.0 mS/cm.

A closer look at concavity of curves reveals that there is a discrepancy of frequencies by a factor of 2 to 3 between curves of corresponding shapes in Figures 5c and 8. A possible reason of this discrepancy is the different waveforms of the ac photocurrent in calculation and experiments. While the calculation assumes an internal current source I_0 producing a sinusoidal waveform with a single frequency, the light beam used in an experiment is turned on and off at a certain frequency, which produces sinusoidal waveforms only at higher frequencies. When the frequency is relatively low, a transient current flows only for a short period after the light beam is turned on or off. The resulting waveform is distorted and contains higher frequency components [13], for which the susceptance becomes larger, and the return current will be larger than calculated.

The waveform of a LAPS current signal can be reproduced by a device simulation, which takes account of the dynamics of minority carriers inside the semiconductor layer [13]. For more precise estimation of the return current, therefore, combination of a circuit model and device simulation should be considered.

7. Conclusions

In this study, a circuit model for the return current in a LAPS was proposed, where the conductivity of the solution and the capacitive coupling between the solution and the semiconductor substrate in the non-illuminated region were formulated as a transmission line. The telegraph equation was numerically solved to find the input admittance of the non-illuminated region, and the dependence of the signal current on various parameters such as the diameter of contact area, the modulation frequency, the specific conductivity of the solution, and the series resistance of the circuit was investigated. It was found that a local change of the analyte concentration in imaging may be underestimated because of the difference of the return current in calibration and in measurement. The dependence of the LAPS signal current on the contact area and the frequency was also observed in experiments, which was compared with the calculated results.

Author Contributions: Conceptualization, T.Y.; methodology, T.Y.; formal analysis, Y.G. and C.F.W.; investigation, D.S. and K.-i.M.; writing—original draft preparation, T.Y.; writing—review and editing, Y.G., C.F.W., and K.-i.M.; project administration, T.Y.

Funding: This work was supported by JSPS KAKENHI Grant Number 17H03074.

Conflicts of Interest: The authors declare no conflict of interest.

References

1. Hafeman, D.G.; Parce, J.W.; McConnell, H.M. Light-addressable potentiometric sensor for biochemical systems. *Science* **1988**, *240*, 1182–1185. [CrossRef] [PubMed]
2. Owicki, J.C.; Bousse, L.J.; Hafeman, D.G.; Kirk, G.L.; Olson, J.D.; Wada, H.G.; Parce, J.W. The light-addressable potentiometric sensor—Principles and biological applications. *Annu. Rev. Biophys. Biomol. Struct.* **1994**, *23*, 87–114. [CrossRef] [PubMed]
3. Yoshinobu, T.; Miyamoto, K.; Werner, C.F.; Poghossian, A.; Wagner, T.; Schöning, M.J. Light-addressable potentiometric sensors for quantitative spatial imaging of chemical species. *Annu. Rev. Anal. Chem.* **2017**, *10*, 225–246. [CrossRef] [PubMed]
4. Nakao, M.; Yoshinobu, T.; Iwasaki, H. Scanning-laser-beam semiconductor pH-imaging sensor. *Sens. Actuators B* **1994**, *20*, 119–123. [CrossRef]
5. Krause, S.; Talabani, H.; Xu, M.; Moritz, W.; Griffiths, J. Scanning photo-induced impedance microscopy—An impedance based imaging technique. *Electrochim. Acta* **2002**, *47*, 2143–2148. [CrossRef]
6. Massobrio, G.; Martinoia, S.; Grattarola, M. Light-addressable chemical sensors: Modelling and computer simulations. *Sens. Actuators B* **1992**, *7*, 484–487. [CrossRef]
7. Sartore, M.; Adami, M.; Nicolini, C. Computer simulation and optimization of a light addressable potentiometric sensor. *Biosens. Bioelectron.* **1992**, *7*, 57–64. [CrossRef]
8. Bousse, L.; Mostarshed, S.; Hafeman, D.; Sartore, M.; Adami, M.; Nicolini, C. Investigation of carrier transport through silicon wafers by photocurrent measurements. *J. Appl. Phys.* **1994**, *75*, 4000–4008. [CrossRef]
9. Sartore, M.; Adami, M.; Nicolini, C.; Bousse, L.; Mostarshed, S.; Hafeman, D. Minority-carrier diffusion length effects on light-addressable potentiometric sensor (LAPS) devices. *Sens. Actuators A* **1992**, *32*, 431–436. [CrossRef]
10. Nakao, M.; Yoshinobu, T.; Iwasaki, H. Improvement of spatial-resolution of a laser-scanning pH-imaging sensor. *Jpn. J. Appl. Phys.* **1994**, *33*, L394–L397. [CrossRef]
11. Parak, W.J.; Hofmann, U.G.; Gaub, H.E.; Owicki, J.C. Lateral resolution of light-addressable potentiometric sensors: An experimental and theoretical investigation. *Sens. Actuators A* **1997**, *63*, 47–57. [CrossRef]
12. George, M.; Parak, W.J.; Gerhardt, I.; Moritz, W.; Kaesen, F.; Geiger, H.; Eisele, I.; Gaub, H.E. Investigation of the spatial resolution of the light-addressable potentiometric sensor. *Sens. Actuators A* **2000**, *86*, 187–196. [CrossRef]

13. Guo, Y.; Miyamoto, K.; Wagner, T.; Schöning, M.J.; Yoshinobu, T. Device simulation of the light-addressable potentiometric sensor for the investigation of the spatial resolution. *Sens. Actuators B* **2014**, *204*, 659–665. [CrossRef]
14. Poghossian, A.; Werner, C.F.; Buniatyan, V.V.; Wagner, T.; Miyamoto, K.; Yoshinobu, T.; Schöning, M.J. Towards addressability of light-addressable potentiometric sensors: Shunting effect of non-illuminated region and cross-talk. *Sens. Actuators B* **2017**, *244*, 1071–1079. [CrossRef]
15. Vogel, Y.B.; Gooding, J.J.; Ciampi, S. Light-addressable electrochemistry at semiconductor electrodes: Redox imaging, mask-free lithography and spatially resolved chemical and biological sensing. *Chem. Soc. Rev.* **2019**, *48*, 3723–3739. [CrossRef] [PubMed]

Article

Estimation of Potential Distribution during Crevice Corrosion through Analysis of *I–V* Curves Obtained by LAPS

Kiyomi Nose [1,*], Ko-ichiro Miyamoto [2] and Tatsuo Yoshinobu [2,3]

1. Material Performance Solution Center, Nippon Steel Technology Corporation, 20-1, Shintomi, Futtu 293-0011, Japan
2. Department of Electronic Engineering, Tohoku University, 6-6, Aza-Aoba, Aramaki, Aoba-ku, Sendai 980-8579, Japan; k-miya@ecei.tohoku.ac.jp (K.-i.M.); nov@ecei.tohoku.ac.jp (T.Y.)
3. Department of Biomedical Engineering, Tohoku University, 6-6, Aza-Aoba, Aramaki, Aoba-ku, Sendai 980-8579, Japan

* Correspondence: nose.kiyomi.58b@nstec.nipponsteel.com

Received: 19 March 2020; Accepted: 14 May 2020; Published: 19 May 2020

Abstract: Crevice corrosion is a type of local corrosion which occurs when a metal surface is confined in a narrow gap on the order of 10 μm filled with a solution. Because of the inaccessible geometry, experimental methods to analyze the inner space of the crevice have been limited. In this study, a light-addressable potentiometric sensor (LAPS) was employed to estimate the potential distribution inside the crevice owing to the IR drop by the anodic current flowing out of the structure. Before crevice corrosion, the *I–V* curve of the LAPS showed a potential shift, depending on the distance from the perimeter. The shift reflected the potential distribution due to the IR drop by the anodic current flowing out of the crevice. After crevice corrosion, the corrosion current increased exponentially, and a local pH change was detected where the corrosion was initiated. A simple model of the IR drop was used to calculate the crevice gap, which was 12 μm—a value close to the previously reported values. Thus, the simultaneous measurement of the *I–V* curves obtained using a LAPS during potentiostatic electrolysis could be applied as a new method for estimating the potential distribution in the crevice.

Keywords: light-addressable potentiometric sensor; LAPS; crevice corrosion; potential distribution; crevice gap

---

1. Introduction

Stainless steel is a highly corrosion-resistant alloy; however, it corrodes under certain circumstances. Crevice corrosion [1–6] is a type of localized corrosion of stainless steel which occurs in the presence of chlorides, where the surface is confined in a narrow gap on the order of 10 μm. It occurs and develops in accordance with the following steps [1,4–6]: (1) consumption of the dissolved oxygen in the crevice, (2) formation of a differential ventilation battery, (3) increase in chloride ion concentration, and (4) lowering of pH owing to a hydrolysis reaction of eluted metal ions. Crevice corrosion depends strongly on the geometry, which affects the transport of ions inside and outside of the crevice.

The crevice corrosion resistance of stainless steel is usually evaluated by determining the critical crevice temperature (CCT) in an immersion test (in accordance with ASTM G48 Method F) [7], measuring the crevice corrosion repassivation potential [8], or conducting an electrochemical corrosion test [9–12] using a potentiostatic test. In electrochemical corrosion tests, an IR drop occurs due to the direct current flowing from the inside of the crevice to the outside. This IR drop becomes larger when the crevice gap is smaller, and when the conductivity of the solution in the crevice is lower. The IR drop results in a potential distribution and therefore a non-uniform biasing condition inside the crevice, depending on the distance from the perimeter. For the potential distribution inside a crevice during a potentiostatic test, a study based on numerical simulation has been reported [13]. However, there have been few experimental studies in which the potential distribution was measured.

In this study, we employed a light-addressable potentiometric sensor (LAPS) [14] to measure the potential distribution inside a crevice. The LAPS is a semiconductor-based chemical sensor in which a light beam generates a photocurrent signal depending on the local potential of the sensor surface at the illuminated position. By using a pH-sensitive Si_3N_4 surface and a scanning light beam, a LAPS can measure the pH distribution of the solution in contact with the sensor surface [15,16]. A LAPS was also applied to the visualization of pH distribution during crevice corrosion [17–19] in a metal/sensor crevice structure formed by mounting a metal specimen directly on the sensor surface. We propose to use the same setup to measure the potential distribution rather than the pH distribution inside a crevice and to estimate the crevice gap, where the IR drop is superimposed on the bias voltage applied to the LAPS.

2. Experimental Methods

2.1. Measurement Setup

Figure 1 shows the configuration of the equipment used for potentiostatic test and the *I–V* curve measurement with the LAPS. The sensor plate was fabricated by depositing the SiO_2 and Si_3N_4 films on an n-type Si substrate, as described in a previous study [16]. A laser beam with a wavelength of 830 nm modulated at a frequency of 2500 Hz illuminated the rear surface of the sensor plate to generate an alternating photocurrent signal (hereinafter expressed as I_p). As the LAPS surface was insulated, only the alternating current flowed through the LAPS sensor plate. Most of the direct current (I_{corr}) returned to the counter electrode (CE).

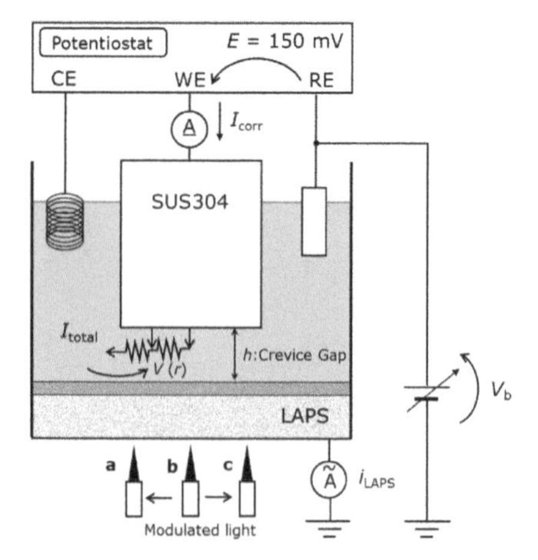

Figure 1. Setup for simultaneously performing the potentiostatic test and light-addressable potentiometric sensor (LAPS) measurement. RE, WE, and CE indicate the reference electrode, working electrode, and counter electrode, respectively. In this experiment, an $Ag/AgCl_{(3M\ NaCl)}$ electrode, a SUS304 specimen, and a platinum wire were used as the RE, WE, and CE, respectively. The potentiostat was used to apply a controlled potential to the specimen and to monitor the corrosion current I_{corr}. The points of illumination **a**, **b**, and **c** correspond to the surface locations shown in the top-right image of Figure 2.

A cylindrical bar made of SUS304 (0.068% carbon, 1.84% manganese, 0.029% phosphorus, 0.027% sulfur, 8.11% nickel, and 18.65% chromium) with a diameter of 12 mm and a height of

50 mm was used as the specimen. This was mounted on the sensor surface to form a metal/sensor crevice structure.

The specimen was subjected in advance to ultrasonic cleaning in acetone and passivated in 30% HNO_3 solution at 50 °C. Two different solutions prepared by diluting artificial seawater (ASW) were used as test solutions. Table 1 lists the specific conductivity and pH of each solution. Here, 5 mL of test solution was poured into the measurement cell, the bottom of which was the sensor surface. To remove the oxide film on the surface of the specimen, the crevice-forming plane of the specimen was polished with #1000 wet sandpaper immediately prior to the experiment. Polishing was completed when the surface did not repel water. The finished surface was mirror-like, which guaranteed fair reproducibility of the corrosion experiment. A large difference in roughness would affect the corrosion due to differences in the surface area and the crevice gap.

Table 1. Specific conductivity and pH of each test solution.

Parameter	1/100 Artificial Seawater	1/10 Artificial Seawater
Specific conductivity (Sm^{-1})	0.074	0.63
pH	6.35	7.21

2.2. *Crevice Corrosion Test*

After the specimen was mounted on the sensor surface, the spontaneous potential of SUS304 was monitored. When the potential value became −200 ± 10 mV, the potentiostatic test and *I–V* curve measurement were started simultaneously. The potentiostatic test was performed at 150 mV vs. $Ag/AgCl_{(3M\ NaCl)}$ (hereinafter expressed by the millivolt value only), and a platinum electrode was used as the CE.

2.3. *I–V Curve Measurement*

During the potentiostatic test, the alternating photocurrent signal I_p of LAPS was repeatedly measured according to the following scheme: first, the bias voltage applied to the LAPS was set at −1400 mV, and the laser beam was moved along the diameter of the SUS304 piece. During this scan, I_p was recorded as a function of the position on the diameter with a constant spacing of 400 µm. The same diameter was repeatedly scanned while changing the bias voltage from −1400 to −1000 mV with an interval of 50 mV, and from −1000 to 0 mV with an interval of 20 mV. After completing these scans, the values of I_p measured at the same position were gathered to construct an *I–V* curve in the range of −1400 to 0 mV, and the bias voltage corresponding to the inflection point of this *I–V* curve was calculated as an indicator of the potential change at that position. This scan was repeated every 145 s throughout the potentiostatic test.

3. Results and Discussion

3.1. *Shift of I–V Curves Obtained by the LAPS during Crevice Corrosion*

The results of a corrosion test in 1/100 ASW are shown in Figure 2. The temporal change of the corrosion current and the *I–V* curves at different positions are presented together with an optical photograph of the corroded surface after the corrosion test. Point **a** is located within the corroded area; point **b** is the center of the specimen, and point **c** is near the right edge of the specimen. Timestamps of 145, 1162, 20,026, 30,031, and 40,036 s are indicated on the curve of the corrosion current, and the *I–V* curves acquired at these points of time are shown below.

The temporal change of the corrosion current shows that the incubation time of crevice corrosion (t_{INCU}) [12] was approximately 1162 s, after which the corrosion current increased exponentially with time. At point **a**, the *I–V* curve moved rightward (i.e., towards higher bias voltages) after the start of the potentiostatic test until t_{INCU}. Then, the *I–V* curve moved leftward with the progress of corrosion.

At point **b**, the rightward shift until t_{INCU} was larger than that of point **a**, and the leftward shift after t_{INCU} was smaller than that of point **a**. At point **c**, the rightward shift was similar to that of point **a**, and the leftward shift after the t_{INCU} was smaller. For interpretation of these results, the shift of the *I–V* curve was quantitatively analyzed as shown in the following sections.

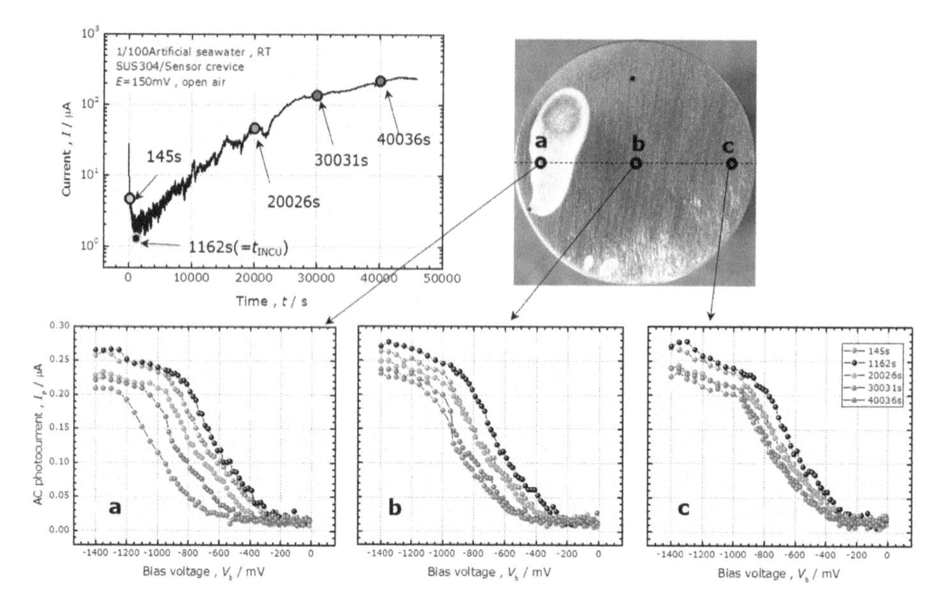

Figure 2. (**Top left**) The temporal change of the corrosion current during the potentiostatic test of a SUS304 specimen at E = 150 mV in 1/100 ASW. (**Top right**) Optical photograph of the corroded surface after 48,132 s of the corrosion test. (**Bottom**) I–V curves measured at different points a, b, and **c**.

3.2. Analysis of the Shift of I–V Curves

To analyze the shift of the *I–V* curves during the potentiostatic test, the inflection point near the middle of the transition region of each *I–V* curve (hereinafter expressed as V_{inf}) was calculated. Figure 3 shows the results for the corrosion in 1/100 ASW. The temporal change of the spatial distribution of V_{inf} is shown separately in Figure 3a,b, for the early stage before t_{INCU} and the later stage after t_{INCU}, respectively.

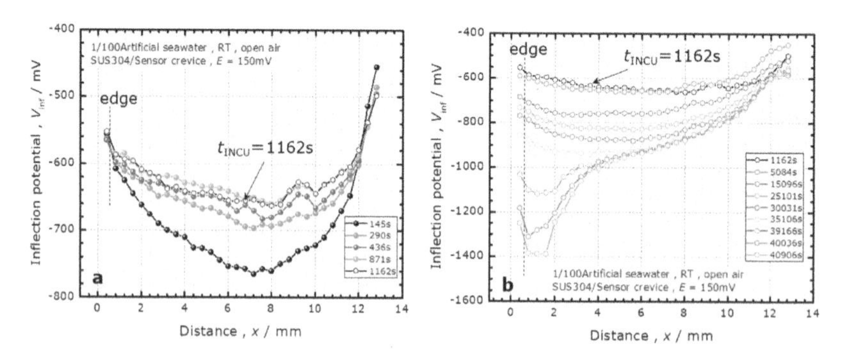

Figure 3. The temporal change of the spatial distribution of the inflection points of *I–V* curves (V_{inf}) in the course of the potentiostatic test at E = 150 mV in 1/100 ASW. (**a**) The early stage before the occurrence of crevice corrosion; (**b**) The later stage after the occurrence of crevice corrosion.

In Figure 3a, the spatial distribution of V_{inf} has a valley shape. V_{inf} decreases as the measurement point goes further from the edge of the specimen and reaches a minimum at approximately the center of the specimen. With increasing time during the potentiostatic test, this valley becomes shallower until the occurrence of corrosion at t_{INCU}. The effective bias applied to the EIS (electrolyte–insulator–semiconductor) structure of a LAPS is the total of the externally applied bias (V_{b}), the change of the Nernst potential sensitive to pH, and the IR drop owing to the DC current flowing out of the crevice. As the pH remains the same, the shift of the *I–V* curve is a direct measure of the IR drop. The pH change remains negligible at the early stage before t_{INCU}; therefore, the spatial distribution of V_{inf} should be attributed to the potential distribution owing to the IR drop. As the corrosion current decreased until t_{INCU}, the IR drop became smaller, and the valley became shallower.

Here, we consider the potential distribution $V(r)$ owing to the IR drop in the crevice. Using the potential outside the crevice as a reference, the $V(r)$ was 0 at the edge of the specimen. When the specimen was anodically polarized, a direct current flowed out of the crevice, and $V(r)$ became higher inside the crevice. In a LAPS measurement, a bias voltage V_{b} is defined as the potential of the $Ag/AgCl_{(3M\ NaCl)}$ reference electrode with respect to the potential of the sensor substrate. Under the potential distribution in the crevice, the effective bias applied to the local position on the sensor surface was $V_{\mathrm{b}} + V(r)$, which explains the spatial distribution of the inflection potential of V_{inf} shown in Figure 3a. The higher the potential $V(r)$, the larger the leftward shift of the *I–V* curve obtained by the LAPS. Simultaneously, when a constant potential of $E = 150$ mV with respect to the reference electrode was applied to the specimen, the effective polarization at the local position on the SUS304 surface was decreased to 150 mV – $V(r)$.

After crevice corrosion at t_{INCU}, the inflection potential V_{inf} decreased, as shown in Figure 3b, suggesting acidification of the solution in the crevice by the reaction:

$$M^{n+} + nH_2O \rightarrow \mathrm{M}(OH)_n + nH^+ \quad (1)$$

where M^{n+} denotes eluted metal ions. Near the left edge of the specimen, where the SUS304 surface was corroded, the V_{inf} sharply decreased after 39,166 s.

3.3. Estimation of the Potential Distribution inside the Crevice

Based on the discussion above, the potential distribution inside the crevice $V(r)$ was equal to the leftward shift of the *I–V* curve, provided that the pH change is negligible. The distribution of the V_{inf} at $t = 145$ s was considered directly after the start of the potentiostatic test. We redefined the position r with respect to the center of the specimen; $r = 0$ is the center, and $r = \pm R$ is the edge of the specimen. If we take the potential at the edge as a reference, $V(r)$ is given by:

$$V(r) = V_{\mathrm{inf}}(R) - V_{\mathrm{inf}}(r) \quad (2)$$

Figure 4 shows the potential distribution $V(r)$ immediately after the start of the potentiostatic test in 1/100 ASW and 1/10 ASW. The specific conductivities of these solutions were 0.074 and 0.63 $\mathrm{Sm^{-1}}$, respectively. In both cases, the potential became higher with increased distance from the edge. The smaller IR drop for 1/10 ASW was consistent with its higher conductivity. The reason for the asymmetry of the curve, especially near the right edge, is unknown; it could be due to a slight tilt of the specimen or convection of the solution.

Figure 4 shows that the value of the $V(r)$ at the center was 149 mV and 60 mV for 1/100 ASW and 1/10 ASW, respectively. When the specimen was biased at $E = +150$ mV vs. $Ag/AgCl_{(3M\ NaCl)}$ reference electrode, the effective polarization voltage at the center of the specimen was estimated to be +1 and +90 mV, respectively. Considering that the spontaneous potential of SUS304 at the start of the potentiostatic test was approximately −200 mV, polarization at +1 and +90 mV was still in the anodic region. Therefore, the entire surface of SUS304 in the crevice was anodically polarized.

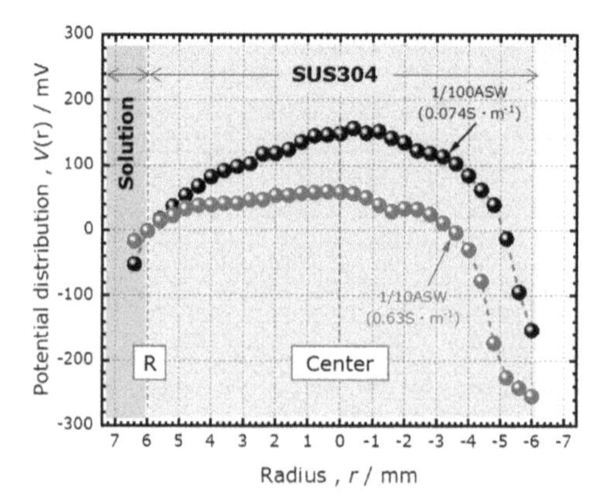

Figure 4. The calculated potential distribution inside the crevice immediately after the start of the potentiostatic test at $E = 150$ mV in 1/100 ASW and 1/10 ASW.

3.4. Estimation of the Crevice Gap

In the crevice corrosion experiment, the SUS304 specimen was placed directly on the sensor surface by its own weight, and the crevice gap remained unknown as it was determined by the surface roughness. Here, we estimated the crevice gap using a simple model, based on the potential distribution obtained in the previous section and the externally measurable total current.

We assumed a uniform crevice gap h and considered a hollow cylinder in the crevice with radius r, thickness dr, and height h, as shown in Figure 5. When the specific conductivity of the solution in the crevice is σ, the resistance between the inner and outer walls of the cylinder filled by the solution is expressed as:

$$\frac{dr}{2\pi rh\sigma} \tag{3}$$

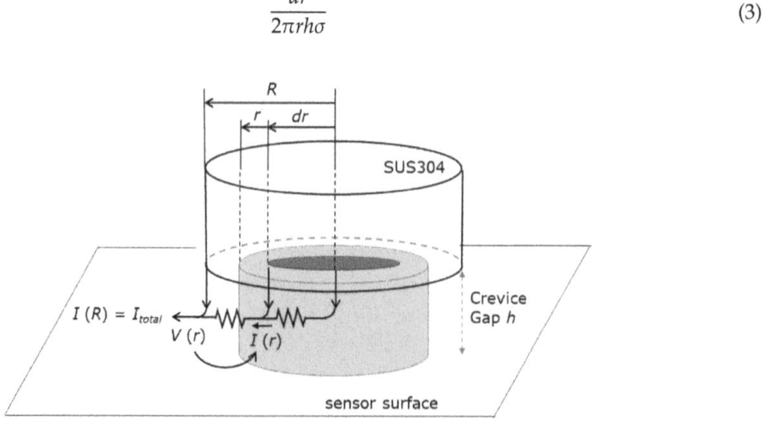

Figure 5. The model used for estimation of the crevice gap.

We define $I(r)$ as the current flowing outward from the outer wall of this hollow cylinder. The IR drop between r and $r + dr$ is given by:

$$dV = -\frac{dr}{2\pi rh\sigma} \cdot I(r). \tag{4}$$

We also define $j(r)$ as the density of the current flowing out of the crevice-forming plane of the specimen into the crevice. The infinitesimal current (dI) flowing into the hollow cylinder is then given by:

$$dI = j(r) \cdot 2\pi r dr. \tag{5}$$

Combining Equations (4) and (5), we obtain:

$$j(r) = \frac{1}{2\pi r} \cdot \frac{dI}{dr} = -h\sigma \left[\frac{d^2V}{dr^2} + \frac{1}{r}\frac{dV}{dr} \right]. \tag{6}$$

By fitting the experimentally obtained $V(r)$ with a parabola,

$$V(r) = A - Br^2, \tag{7}$$

we can determine the fitting parameters A and B, which will give:

$$\frac{d^2V}{dr^2} + \frac{1}{r}\frac{dV}{dr} = -4B. \tag{8}$$

Therefore, an approximation of $V(r)$ with a parabola implies that the current density in Equation (6) is independent of r, and is given by:

$$j = 4Bh\sigma. \tag{9}$$

The polarization current I_{total} (passive current) flowing into the crevice is then given by:

$$I_{total} = 4Bh\sigma\pi R^2, \tag{10}$$

which can be used to determine the value of h. We applied this analysis to the experimentally obtained potential distribution $V(r)$ directly after the start of the potentiostatic test in 1/100 ASW, which is shown in Figure 4. Fitting this curve with a parabola as shown in Figure 6, the values of A and B were determined to be 0.147 V and 4140 Vm^{-2}, respectively. As shown in the top-left image of Figure 2, the corrosion current measured at 145 s was 4.6 μA, which included the current flowing into the crevice (I_{total}) and the current flowing out of the side wall of the specimen contacting the solution. Considering the ratio of the surface areas at the bottom and on the side wall of the specimen, where SUS304 contacted the solution, I_{total} was estimated to be 1.7 μA. Using these values as well as R = 6 mm and σ = 0.074 Sm^{-1} in Equation (10), the crevice gap h was estimated to be 12 μm. This value is similar to the previously reported values of 6 to 12 μm obtained by measuring the weight of ethanol filling the crevice gap between an SUS304 surface and a quartz glass rod [11].

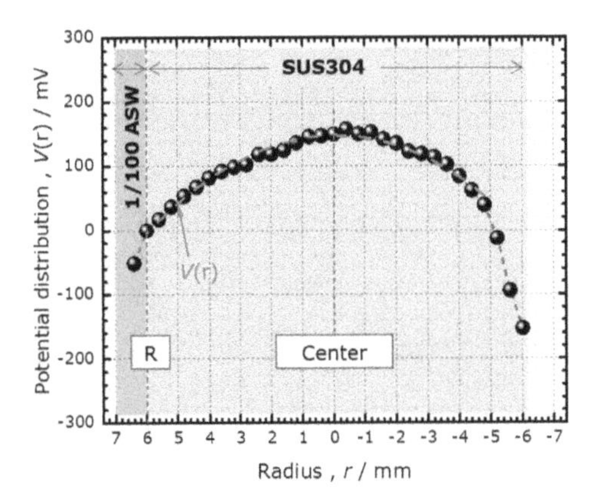

Figure 6. The fitting of the experimentally obtained potential distribution $V(r)$ with a parabola.

4. Conclusions

In this study, a LAPS was employed to analyze the potential distribution inside a narrow gap during crevice corrosion of SUS304. A potentiostatic test and *I–V* curve measurement by a LAPS were simultaneously performed. The setup included an SUS304 specimen placed directly on the LAPS surface with a narrow gap filled with ASW. Before the crevice corrosion at t_{INCU}, the plot of the inflection point of the *I–V* curve showed a valley shape across the crevice, reflecting the potential distribution due to the IR drop by the anodic current flowing out of the crevice. After t_{INCU}, the corrosion current increased exponentially, and a local change of the inflection point due to the lowering pH was observed at the position where the surface was corroded. A simple model of the IR drop was proposed to describe the potential distribution before t_{INCU}, which was used to estimate the crevice gap by fitting the experimental data. The estimated gap of 12 μm was similar to the previously reported values obtained by measuring the weight of the liquid filling the gap.

Author Contributions: Conceptualization, K.N.; methodology, K.N.; formal analysis, K.N. and T.Y.; investigation, K.N. and K.-i.M.; writing—original draft preparation, K.N.; writing—review and editing, T.Y.; project administration, T.Y. All authors have read and agreed to the published version of the manuscript.

Conflicts of Interest: The authors declare no conflict of interest.

References

1. Oldfield, J.W. Test techniques for pitting and crevice corrosion resistance of stainless steels and nickel-base alloys in chloride-containing environments. *Int. Mater. Rev.* **1987**, *32*, 153–172. [CrossRef]
2. Hisamatsu, Y. Localized corrosion of stainless steels Part1 Pitting and crevice corrosion. *Testu-to-Hagane* **1977**, *5*, 574–584. [CrossRef]
3. Suzuki, T. Localized corrosion. *Bosyoku-Gijyutu* **1976**, *25*, 761–768.
4. Suzuki, T. Crevice corrosion of stainless steel. *Bosyoku-Gijyutu* **1979**, *28*, 38–45.
5. Gösta, K.; Gösta, W. On the mechanism of crevice corrosion of stainless Cr steels. *Corros. Sci.* **1971**, *11*, 499–510.
6. Dayal, R.K. *Crevice Corrosion of Stainless Steel. Corrosion of Austenitic Stainless Steels*, 1st ed.; Woodhead Publishing: Sawston, UK, 2002; Chapter 4, pp. 106–116.
7. ASTM International, ASTM G48-11: 2015. *Standard Test Methods for Pitting and Crevice Corrosion Resistance of Stainless Steels and Related Alloys by Use of Ferric Chloride Solution*; ASTM International: West Conshohocken, PA, USA, 2015.
8. Japanese Industrial Standards, JIS G 0592:2002. *Method of Determining the Repassivation Potential for Crevice Corrosion of Stainless Steels*; Japanese Standards Associa: Tokyo, Japan, 2002.
9. Suzuki, T.; Yoshihara, K. Critical potential for growth of localized corrosion of stainless steel in chloride media. *Corros. NACE* **1972**, *28*, 1–6. [CrossRef]
10. Nishimoto, M.; Ogawa, J.; Muto, I.; Sugawara, Y.; Hara, N. Simultaneous visualization of pH and Cl^- distributions inside the crevice of stainless steel. *Corros. Sci.* **2016**, *106*, 298–302. [CrossRef]
11. Mastuhashi, R.; Matsuoka, K.; Nose, K.; Kajimura, H. Propagation analysis of crevice corrosion on SUS304 stainless steel under potentiostatic condition in artificial sea water. *Zairyo-to-Kankyo* **2015**, *64*, 51–59. [CrossRef]
12. Sakiya, M.; Matsuhashi, R.; Matsuhashi, T.; Takahashi, A. The effect of potential and temperature on crevice corrosion incubation time for stainless steels in diluted chloride ion environment. *Zairyo-to-Kankyo* **2009**, *58*, 378–385. [CrossRef]
13. Mastuoka, K.; Matsuhashi, R.; Nose, K.; Kajimura, H. The numerical analysis potential/current density distribution in crevice corrosion propagation stage. *Zairyo-to-Kankyo* **2016**, *65*, 350–357.
14. Hafeman, D.G.; Wallace Parce, J.; McConnell, H.M. Light-addressable potentiometric sensor for biochemical systems. *Science* **1988**, *240*, 1182–1185. [CrossRef] [PubMed]
15. Nakao, M.; Yoshinobu, T.; Iwasaki, H. Scanning-laser-beam semiconductor pH-imaging sensor. *Sens. Actuators B* **1994**, *20*, 119–123. [CrossRef]

16. Yoshinobu, T.; Miyamoto, K.; Werner, C.F.; Poghossian, A.; Wagner, T.; Schöning, M.J. Light-addressable potentiometric sensors for quantitative spatial imaging of chemical imaging of chemical species. *Annu. Rev. Anal. Chem.* **2017**, *10*, 225–246. [CrossRef] [PubMed]
17. Miyamoto, K.; Sakakita, S.; Wagner, T.; Schöning, M.J.; Yoshinobu, T. Application of chemical imaging sensor to in-situ pH imaging in the vicinity of a corroding metal surface. *Electrochim. Acta* **2015**, *183*, 137–142. [CrossRef]
18. Miyamoto, K.; Sakakita, S.; Werner, C.F.; Yoshinobu, T. A modified chemical imaging sensor system for real-time pH imaging of accelerated crevice corrosion of stainless steel. *Phys. Status Solidi A* **2018**, *215*, 1700963. [CrossRef]
19. Nose, K.; Kajimura, H.; Miyamoto, K.; Yoshinobu, T. The pH in crevice measured by a semiconductor chemical sensor and relationshisp with crevice corrosion behavior of stainless steel. *Zairyo-to-Kankyo* **2020**, *69*, 40–48.

Article

A LAPS-Based Differential Sensor for Parallelized Metabolism Monitoring of Various Bacteria

Shahriar Dantism [1,2], Désirée Röhlen [1], Torsten Wagner [1,3], Patrick Wagner [2] and Michael J. Schöning [1,3,*]

1 Institute of Nano- and Biotechnologies (INB), FH Aachen, Heinrich-Mußmann-Straße 1, 52428 Jülich, Germany; dantism@fh-aachen.de (S.D.); roehlen@fh-aachen.de (D.R.); torsten.wagner@fh-aachen.de (T.W.)
2 Department of Physics and Astronomy, Laboratory for Soft Matter and Biophysics, KU Leuven, Celestijnenlaan 200 D, 3001 Leuven, Belgium; patrickhermann.wagner@kuleuven.be
3 Institute of Complex Systems (ICS-8), Research Centre Jülich GmbH, Wilhelm-Johnen-Straße 1, 52425 Jülich, Germany
* Correspondence: schoening@fh-aachen.de; Tel.: +49-241-6009-53215

Received: 10 October 2019; Accepted: 24 October 2019; Published: 29 October 2019

Abstract: Monitoring the cellular metabolism of bacteria in (bio)fermentation processes is crucial to control and steer them, and to prevent undesired disturbances linked to metabolically inactive microorganisms. In this context, cell-based biosensors can play an important role to improve the quality and increase the yield of such processes. This work describes the simultaneous analysis of the metabolic behavior of three different types of bacteria by means of a differential light-addressable potentiometric sensor (LAPS) set-up. The study includes *Lactobacillus brevis*, *Corynebacterium glutamicum*, and *Escherichia coli*, which are often applied in fermentation processes in bioreactors. Differential measurements were carried out to compensate undesirable influences such as sensor signal drift, and pH value variation during the measurements. Furthermore, calibration curves of the cellular metabolism were established as a function of the glucose concentration or cell number variation with all three model microorganisms. In this context, simultaneous (bio)sensing with the multi-organism LAPS-based set-up can open new possibilities for a cost-effective, rapid detection of the extracellular acidification of bacteria on a single sensor chip. It can be applied to evaluate the metabolic response of bacteria populations in a (bio)fermentation process, for instance, in the biogas fermentation process.

Keywords: light-addressable potentiometric sensor (LAPS); *Lactobacillus brevis*; *Escherichia coli*; *Corynebacterium glutamicum*; cellular metabolism; differential cell-based measurement; multi-analyte analysis; extracellular acidification

---

1. Introduction

Food digestion within the gastrointestinal tract is a good example for biofermentation processes in our daily life, in which many different types of microorganisms can be involved. Without microbes, large and complex food molecules cannot be broken down into required nutrients for the human body [1]. In the food-research sector, complex microbial interactions have been studied to enhance, for instance, the aroma profile and flavor in soy sauce during the fermentation process [2]. As another example, microorganisms play a major role in coffee fermentation by degrading the mucilage to alcohols, acids, and enzymes [3]. In agricultural biogas plants, various types of bacteria contribute to the conversion of biomass (e.g., maize silage) into a usable energy source (e.g., methane gas) [4]. In all applications of bioreactor technology, the on-line monitoring of the metabolic activity of microorganisms should be seriously considered to avoid undesired, time-consuming, and cost-intensive interventions, which can reduce the yield at the end of the production chain. Related fields of application include: cell health monitoring in bioreactors [5], continuous non-invasive monitoring of cell growth in disposable bioreactors [6], on-line near-infrared bioreactor monitoring of cell density [7], online monitoring of cell concentration in high-cell density *Escherichia coli* cultivations [8], sensing metabolites for monitoring the tissue-engineered cellularity in perfusion bioreactors [9], and micro-biosensors for fed-batch fermentation with integrated online monitoring [10]. In all of those examples, analytical sensors are the enabling element for rapid, sensitive, and cost-effective detection of various parameters (e.g., health, growth, and density of cells in bioreactors). In this context, light-addressable potentiometric sensors (LAPS) can be applied as suitable tools for monitoring the extracellular acidification of cells. A LAPS is a field-effect-based chemical sensor with the ability to monitor concentration changes of biochemical/biological species in a spatially resolved way [11–13]. It belongs to the family of electrolyte/insulator/semiconductor (EIS)-based capacitive sensors [14]. By addressing defined regions of interest on a sensor chip with modulated light beams such as laser-diode modules, two-dimensional (2D) chemical images of concentration distributions can be recorded [15–18]. In comparison to other 2D potentiometric chemical imaging sensors applying e.g., arrays of ion-sensitive field-effect transistors (ISFETs) [19] or charge-coupled devices (CCDs) [20], LAPS require no sensor patterns to record chemical images: The LAPS surface does not require pattering, wiring or passivation, which allows bacteria to come directly into contact with the pH-sensitive transducer layer to determine the extracellular acidification. Furthermore, on its planar sensor surface, multi-chamber structures can be attached to perform differential measurements with distinct cell suspensions [21]. The principle of differential measurements allows elimination of unwanted external influences such as sensor signal drift and pH value variations of the measurement solution during experiments [22]. Beside the chemical imaging technique visualizing the pH distribution on the sensor surface, there are a variety of further LAPS sensing techniques such as the scanned light-pulse technique (SLPT) in studies of interface properties (e.g., flat-band voltage) [15], or scanning photo-induced impedance microscopy (SPIM) analyzing impedance changes e.g., in PAH/PSS polyelectrolyte microcapsules labeled with Au nanoparticles [23]. The equivalent circuit diagram for designing the LAPS set-up consists of three key parts (illumination area, non-illumination area, and an external circuit), which are described in details in [15]. The sensor fabrication steps are described in Section 2.1. Explanations about LAPS operation modes (e.g., constant-bias, constant-current, potential-tracking, and phase-mode) can be found in [24].

A few examples of on-going research for LAPS-based biosensing are the determination of the extracellular acidification of *Escherichia coli* [25–27], and differential imaging of the metabolism of bacteria and eukaryotic cells [28]. In this regard, quantitative differential monitoring of the metabolic activity of *Corynebacterium glutamicum* [29], and image detection of yeast *Saccharomyces cerevisiae* [30] have been recently discussed. Correspondingly, dual functional extracellular recording for better signal transduction [31], and monitoring secretion of adrenal chromaffin cells by local extracellular acidification [32] are further applications.

In this work and for the first time, the extracellular acidification of three model microorganisms, namely *Escherichia coli* (*E. coli*) K12, *Corynebacterium glutamicum* (*C. glutamicum*) ATCC13032, and *Lactobacillus brevis* (*L. brevis*) ATCC 14869 is assessed simultaneously by means of a four-chamber differential LAPS set-up by varying the cell number and/or glucose concentration. These model microorganisms were selected, as acid-forming, facultative anaerobe, easy-to-cultivate, and commonly used bacteria in laboratory and industrial applications [25,33,34]. Both *E. coli* and *C. glutamicum* have been studied in separate LAPS experiments recently [27,29], however, the cellular metabolism of *L. brevis* has not been analyzed by LAPS so far. In a first experimental step, the extracellular acidification of *L. brevis* cells has been evaluated: calibration curves were established, which render the potential-change rate as a function of glucose concentration and cell number. Data were compared with results of already in literature discussed microorganisms such as *E. coli* and *C. glutamicum*. Finally, a novel parallelized measurement procedure is introduced, which allows sequentially and simultaneously performed LAPS measurements with all three types of bacteria.

The motivation is briefly explained as follows: In terms of the specific metabolic characteristic of each model microorganism, distinguishable sensor signal responses with LAPS can be obtained.

This way, different signal patterns of studied bacteria can be saved in a database. Later on, such data can be applied as references to evaluate the cellular metabolism of bacteria populations in a fermentation process (e.g., in biogas processes). Here, the mutual metabolic influence of cells in the fermenter broth on the metabolization of model microorganisms can be studied. Hence, signal variations after 'interactions' between bacteria can be detected. This approach can contribute to a better understanding of the fermentation process and help to avoid bacteria-related process crashes in a bioreactor. In addition, the multi-analyte differential measurement on a single LAPS chip underlines the possibility of combinatorial analysis with different cell types in parallel enabling a fast data collection utilizing a capacitive field-effect biosensor.

2. Materials and Methods

2.1. Sensor Fabrication and Measurement Set-Up

The LAPS chip consists of an Al/p-Si/SiO_2/Ta_2O_5 field-effect structure: Starting with a p-doped silicon wafer (<100>, 5–10 Ωcm, thickness: 540 μm), 30 nm SiO_2 were grown by a thermal dry oxidation step (O_2, 40 min at 1000 °C), followed by deposition of the Ta_2O_5 layer (electron-beam evaporation (0.5 nm/s, at 6×10^{-6} mbar) of 30 nm Ta and subsequent oxidation (45 min at 520 °C) to 60 nm Ta_2O_5). The ohmic rear-side contact of 300 nm Al on Si was also prepared by electron-beam evaporation. Subsequently, the silicon wafer was diced into single LAPS chips of 20×20 mm^2 size, and a part of the rear-side contact was removed by wet-chemical etching (5% hydrofluoric acid) defining a window for the rear-side illumination. The sensor fabrication steps are described in detail in the references [35–38]. The LAPS chip was mounted in a home-made measurement cell and the set-up is schematically illustrated in Figure 1.

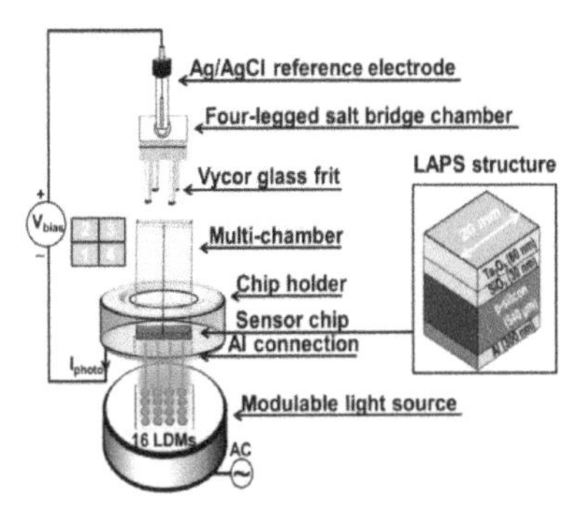

Figure 1. Schematic illustration of the four-chamber differential LAPS measurement set-up, consisting of an Ag/AgCl reference electrode, a PP-ABS four-legged salt bridge chamber (total height 40 mm, leg height 20 mm, $Ø_{legs}$ 4 mm, container area = 15×15 mm^2) combined with four Vycor glass frits ($Ø_{frit}$ 3.2 mm, height 4 mm) filled with 1 mL KCl solution (3 M), a PP-ABS-based four-chamber structure (sensing area per chamber ≈ 7×7 mm^2, height 20 mm), a chip holder (Ø 40 mm, height 10 mm) made of PEEK, a LAPS chip with Al/p-Si/SiO_2/Ta_2O_5 layers, and a light source based on an array containing 16 infrared laser-diode modules (4 per chamber, Ø 3.3 mm, length 7 mm). Chambers 1, 2, and 3 are used as active sensor side with cells. Chamber 4 serves as a reference chamber without cells. The active rear-side illumination area is 15×15 mm^2. I_{photo}: photocurrent, V_{bias}: bias voltage, LDMs: laser-diode modules, AC: alternating current.

The measurement cell consists of a 3D-printed photopolymer-based (polypropylene-acrylonitrile-butadiene-styrene, PP-ABS) four-chamber structure combined with a four-legged salt bridge chamber (filled with 1 mL of 3 M KCl solution) and a polymer-based (polyether ether ketone, PEEK) chip holder, housing the LAPS chip as working electrode and a commercial Ag/AgCl (Metrohm GmbH) reference electrode. More information related to the design, size, and performance of all constructed polymer structures can be found in previous publications [19,27,28]. The four-chamber structure enables to study different cell types simultaneously. In the following experiment, three chambers (1–3) serve as active sensor site with various cell suspensions, while the fourth chamber was used as a reference chamber without cells. This way, pH- and temperature fluctuations as well as a signal drift of the sensor can be compensated. To read out the LAPS sensor signal, a DC (direct current) bias voltage (V_{bias}) is applied between the Ag/AgCl reference electrode and the rear-side contact. The illumination unit is based on an array of 16 small-sized, fixed-focus, and tunable infrared laser-diode modules (LDMs, λ = 785 nm, Roithner Lasertechnik GmbH, Vienna, Austria, serial number APCD-780-07-C3). A simultaneous modulation with different frequencies (frequency divider with a constant value of 160, main clock: 160 MHz, sampling frequency: 1 MHz) was carried out through a field-programmable gate array (FPGA)-based microcontroller, see details in [39]; the illuminated and non-illuminated sensor areas can be modeled with equivalent circuit diagrams as discussed in [40].

The bias voltage V_{bias} allows to induce the formation of a space-charge region within the p-doped silicon at the insulator/semiconductor interface [41]. Here, the charge carriers, i.e., electron-hole pairs induced by the light source, are separated in the electrical field, resulting in a photocurrent, I_{photo}. This photocurrent depends on the surface potential of the LAPS chip. Due to the direct contact of the pH-sensitive Ta_2O_5 transducer layer with the microorganisms in the analyte (chambers 1, 2, and 3), changes in their metabolic activity (extracellular acidification) will consequently lead to changes in the surface potential through a variation of the H^+-ion activity of the LAPS surface [21]. Besides of current–voltage (I–V) measurements, the LAPS chips were electrochemically characterized by capacitance–voltage- (C–V), impedance spectroscopy-, leakage-current-, and constant-capacitance (ConCap) measurements utilizing an electrochemical spectrum analyzer (Zahner-Elektrik GmbH). The sensitivity of the transducer structure (54 mV/pH) was determined as described in [27]. Further information about the electrochemical characterization of the LAPS chips and the FPGA-based set-up can be also found in [42–45].

2.2. *Sample Preparation and Microorganism Cultivation*

For the preparation of glucose solutions and cell suspensions, diluted phosphate-buffered saline (PBS) was used as stock solution: Here, 0.2 g of KCl, 8 g of NaCl, 1.15 g of Na_2HPO_4, and 0.2 g of KH_2PO_4 were dissolved in 1 L distilled water. The buffer solution was further diluted with distilled water, so that a total buffer capacity of 0.2 mM was obtained. A low-buffer capacity is required to observe the extracellular acidification of cells on the LAPS. To determine the buffer capacity, a titration method was used with hydrochloric acid (HCl, 1 M). The buffer was autoclaved at 121 °C for 2 h. Subsequently, the pH value was adjusted at pH 7.4 using NaOH/HCl (1 M) solution. Different glucose concentrations (1.67, 2.5, 3.33, and 5 mM) were prepared after a serial dilution of the stock glucose solution (10 mM) utilizing the diluted PBS. The procedure to cultivate Gram-negative, rod-shaped *E. coli* K12 bacteria (24×10^9 CFU/mL cells) is described in [21,27]. Gram-positive, rod-shaped *C. glutamicum* ATCC13032 bacteria were cultivated on a *Corynebacterium* agar consisting of 10 g of casein peptone, 5 g of yeast extract, 5 g of glucose, 5 g of NaCl, 15 g of agar, in 1 L distilled water. First, cells were incubated at 30 °C by 141 rpm for about 6 h. Optical density measurements (Fisher Scientific, GE Healthcare Ultraspec^{TM} 2100 pro) were used to determine the cell growth density (λ = 578 nm) in all cell suspensions. The overnight cultivation was performed in an incubator around 12 h at 30 °C. This way, 24×10^9 CFU/mL cells were harvested after two washing steps with the diluted PBS solution. After the centrifugation step, the cell pellet was resuspended in the buffer solution (pH 7.4, 0.2 mM).

For the cultivation of Gram-positive, heterofermentive, rod-shaped *L. brevis* ATCC 14869 bacteria, MRS- (De Man, Rogosa and Sharpe) selective culture medium was applied. The MRS agar contains 10 g of casein, 10 g of meat extract, 5 g of yeast extract, 20 g of glucose, 1 g of TWEEN 80, 2 g of K_2HPO_4, 5 g of Na-acetate, 0.2 g $MgSO_4 \times 7\ H_2O$, 0.05 g of $MNSO_4 \times H_2O$, and 1 L of distilled water. The pH value was adjusted to pH 6.2. Cells were incubated at 30 °C by for about 48 h. Further steps are similar to the cultivation of *C. glutamicum* cells, as mentioned above to obtain 24×10^9 CFU/mL cells.

With all cultivated microorganisms, five different cell numbers (0.3×10^9, 0.6×10^9, 1.2×10^9, 2.4×10^9, 4.8×10^9 cells in 200 µL suspension) were applied for the four-chamber differential LAPS measurements by a 1:2 dilution series with PBS solution to study their metabolic behavior. The as-prepared cell suspensions were used on the same day, in which LAPS measurements were performed. Freshly cultured suspensions were necessary for each measurement day to guarantee the reproducibility of the experiments. All final pH values of the cell suspensions were adjusted to 7.4 before starting with the cell-based differential LAPS measurements and controlled by a conventional pH-glass electrode (type: DGi115-SC, Mettler Toledo, Zurich, Switzerland).

3. Results and Discussion

3.1. Monitoring the Cellular Metabolism of L. Brevis Bacteria

The analysis of the metabolic activity of *E. coli* and *C. glutamicum* has been recently reported in [27,29]. This section describes the determination of cellular metabolism of *L. brevis* cell suspensions utilizing the four-chamber differential LAPS set-up. Three different cell numbers (1.2×10^9, 2.4×10^9, 4.8×10^9 cells) were chosen to monitor the acidification behavior. In the first measurement (Figure 2, at 1.67 mM), all chambers were loaded with 100 µL glucose (end resulting concentration: 1.67 mM, pH 7.4). The sensor chip was first conditioned with the glucose solution for 10 min, then 200 µL of cell suspension (pH 7.4) were added into chambers 1, 2, and 3 and 200 µL of the diluted PBS solution in the reference chamber 4 (see set-up in Figure 1). Thus, a total suspension volume per chamber of 300 µL was used. For each chamber, four light spots in the area of interest were selected to read-out the sensor signal. After that, the mean values of the recorded potential change values per chamber were plotted (see diagram I). This measurement procedure was repeated for three further glucose concentrations (2.5, 3.33, and 5 mM) shown in the diagrams II, III, and IV.

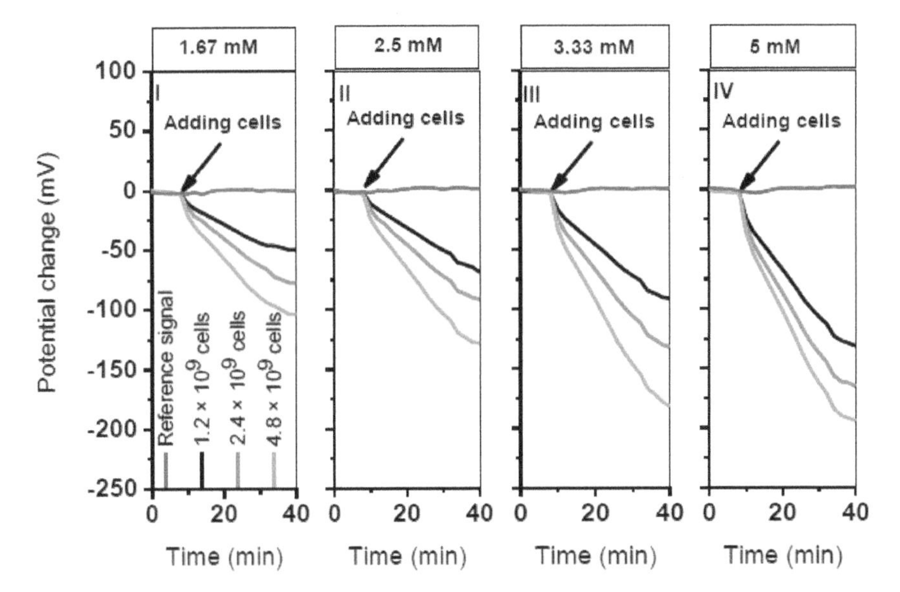

Figure 2. Four-chamber differential LAPS measurement with varying cell number of *L. brevis* (1.2×10^9, 2.4×10^9, 4.8×10^9 cells in 200 µL cell suspension) and varying glucose concentration (1.67, 2.5, 3.33, 5 mM). Potential changes of four successive independent measurements with an ascending series of glucose concentrations. Blue line: reference sensor signal without cells; black, red, and green lines: active sensor sites with cells. Four laser-diode modules (LDMs) were considered for each chamber (each curve corresponds to calculated mean values).

Figure 2 depicts the mean values of the potential changes *vs.* time in four independent successive measurements performed with increasing both the glucose concentration and the cell number. After each independent measurement of 40 min (first, the sensor chip was conditioned for 10 min without cells; then, the respective cell concentration was added), the sensor chip was washed with the diluted PBS solution (pH 7.4, 0.2 mM) and the next glucose concentration (e.g., 2.5 mM) was pipetted into the sensor chambers. Four output signals from four chambers are marked in different colors: in the first measurement at 1.67 mM glucose (diagram I), the blue line indicates the reference signal in the absence of cells in chamber 4. The black line refers to the cell suspension with the lowest cell number of 1.2×10^9 cells, which results in a potential drop of about 50 mV. The signal from the chamber with 2.4×10^9 cells is shown in red and indicates a potential change of approximately 77 mV. The highest signal change of approximately 103 mV with 4.8×10^9 cells in chamber 3 is plotted in green, which corresponds to a pH value of ca. 5.5 on the sensor surface when considering a pH sensitivity of 54 mV/pH of the LAPS chip without cell suspensions [27]. In the measurement with the highest glucose concentration of 5 mM, the highest potential drop of approximately 194 mV was observed again in chamber 3 (green line) with 4.8×10^9 cells, which corresponds to a pH shift at the LAPS surface of $\Delta pH \approx 3.8$. Further potential change values of 131 mV and 164 mV were achieved for cell numbers 1.2×10^9 and 2.4×10^9 cells in chamber 1 and chamber 2, respectively. For all measurements, the metabolic response of the acid-forming *L. brevis* bacteria induces potential changes after adding cells, which leads to an increase of H^+-ion activity on the transducer surface. The extracellular acidification due to the metabolic activity causes a shift to the negative V_{bias} axis and a respective potential drop. The experiments have shown that by increasing the cell number and/or the glucose concentration in each measurement, higher potential change values can be achieved, as a result of metabolically active microorganisms and comparable to results obtained with *E. coli* and *C. glutamicum*, respectively [27,29]. From Figure 2, differential signals can be evaluated by subtracting the signal values of chambers with cells from the reference chamber without cells. The slope of the decreasing differential signals can be calculated in a particular time period (here, within the first 6 min after adding cells, i.e., from minute 10 to 16). These values describe the potential change rates (PCR) given in mV/min and can be computed through a linear regression method. Figure 3 represents the corresponding 3D plot of the calculated PCR values for different cell numbers (1.2×10^9, 2.4×10^9, and 4.8×10^9 cells) and glucose concentrations (1.67, 2.5, 3.33, and 5 mM). For the detailed values, see Table S1 in supplementary information of this article.

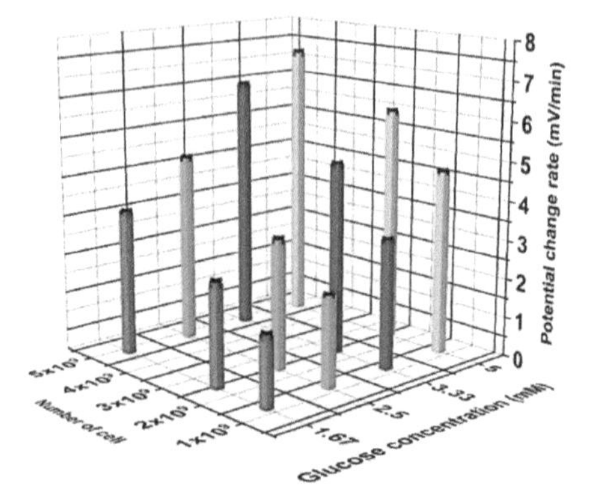

Figure 3. 3D plot of mean values and standard deviations (three repetitions) of the potential change rates for four successive independent measurements with different glucose concentrations (1.67, 2.5, 3.33, 5 mM) when varying the cell numbers (1.2×10^9, 2.4×10^9, 4.8×10^9 cells) in 200 µL of *L. brevis* cell suspension. Data are calculated from Figure 2. Mean values and standard deviations of PCR values were obtained from three independent measurement repetitions within the first 6 min after adding cells. For each chamber, four laser-diode modules (LDMs) were used. The mean value of signals per chamber was considered for further evaluations.

By increasing the cell number and/or glucose concentration, the extracellular acidification and the PCR values increase. For the lowest cell number of 1.2×10^9 cells at 1.67 mM glucose, the lowest PCR value of 1.76 ± 0.05 mV/min was calculated. At the highest applied glucose concentration of 5 mM and the highest cell number of 4.8×10^9 cells, the highest LAPS signal response of 7.20 ± 0.05 mV/min was recorded.

3.2. Determination of Calibration Curves for L. brevis, C. Glutamicum, and E. Coli

In this section, the PCR values of three microorganisms, namely *L. brevis*, *C. glutamicum*, and *E. coli*, are compared with each other by means of calibration curves as a function of glucose concentration or cell number. Different types of bacteria induce different independent LAPS signal responses due to their diverse acidification behavior on the sensor surface. With the knowledge of their signal characteristics (calibration curves), these model microorganisms might be used to analyze metabolic responses of microorganism populations within a (bio)fermentation broth (e.g., from a biogas reactor). It can be studied, whether/how cells in the fermentation broth from different process stages will influence the extracellular acidification of studied model bacteria: calibration matrices can be defined that later-on can be correlated with different scenarios of the biogas process. In this way, an external feedback control of the biogas operation might be envisaged. In our experiments, two approaches were separately considered: First, glucose concentrations were varied at a constant cell number (4.8×10^9 cells) to find a correlation between the sensor signal and the glucose uptake. Second, calibration curves were defined by varying the particular cell number at a constant glucose concentration (1.67 mM). For the determination of the PCR values of the cellular metabolism of *E. coli* and *C. glutamicum*, the same procedure was performed with the differential LAPS set-up, as described in Section 3.1 and shown in Figure 2. Figure 4 depicts the corresponding calibration curves with the PCR mean values for variations of the glucose concentration between 0.042 mM and 5 mM for the three studied microorganisms at a constant cell number of 4.8×10^9 cells. For the detailed values, see Table S2 in supplementary information of this article.

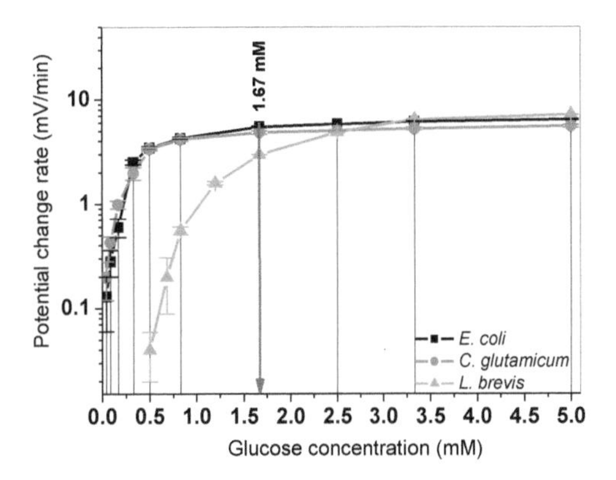

Figure 4. Correlations between the LAPS signal response and the glucose concentration. Mean values and standard deviations of PCR values of *E. coli* (black), *C. glutamicum* (red), and *L. brevis* (green) are depicted for different glucose concentrations (0.042, 0.085, 0.17, 0.20, 0.33, 0.40, 0.50, 0.68, 0.83, 1.20, 1.67, 2.50, 3.33, and 5 mM) at a constant cell number of 4.8×10^9 cells. Three independent measurements were performed within the first 6 min after adding cells to calculate the mean PCR values. The blue arrow in the diagram indicates a particular glucose concentration (1.67 mM), which was exemplarily chosen to compare calibration curves of metabolic responses of the three microorganisms by cell number variations, see also Figure 5.

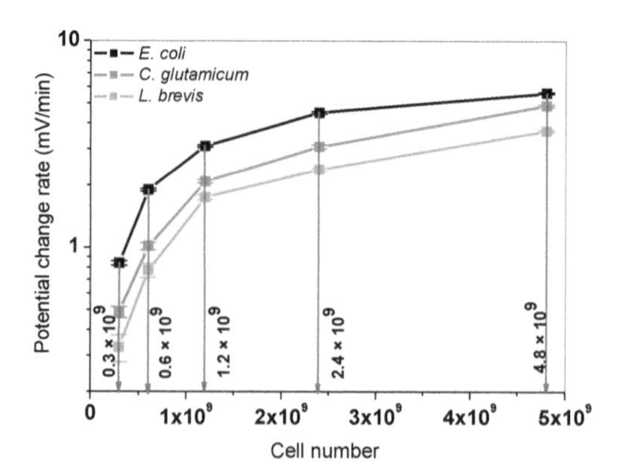

Figure 5. Correlation between the LAPS signal response and the cell number. The mean values and standard deviations of PCR values of *E. coli* (black), *C. glutamicum* (red), and *L. brevis* (green) are depicted for different cell numbers (0.3×10^9, 0.6×10^9, 1.2×10^9, 2.4×10^9, 4.8×10^9 cells) at a constant glucose concentration of 1.67 mM. Three independent measurement repetitions were performed.

In Figure 4, the extracellular acidification of *E. coli* and *C. glutamicum* increases after increasing the glucose concentration, comparable with the results in [27,29]. For *E. coli* and *C. glutamicum* bacteria, the lowest detectable signal change was found for 0.042 mM glucose and for *L. brevis* it was 0.5 mM both at a constant cell number of 4.8×10^9 cells. In experiments with *E. coli* bacteria, the PCR values increased from 0.13 ± 0.07 mV/min to 6.40 ± 0.10 mV/min upon increasing glucose concentrations from 0.042 mM to 5.0 mM. The PCR values of *C. glutamicum* bacteria increased from 0.20 ± 0.08 mV/min to 5.60 ± 0.15 mV/min after raising glucose concentrations, respectively (similar to glucose concentrations for *E. coli* in Figure 4). However, the first detectable PCR value 0.04 ± 0.01 mV/ min for *L. brevis* cells could be obtained at a glucose concentration of 0.50 mM. For lower glucose concentrations (e.g., from 0.042 to 0.4 mM), no detectable signal changes could be monitored. It seems to be that *L. brevis* bacteria require more glucose to be able to form enough H^+-ions on the LAPS surface (shifted green curve in Figure 4). That is supported by the requirement of a higher amount of glucose (20 g/L) during the cultivation phase compared to *E. coli* (1 g/L) and *C. glutamicum* (5 g/L). The glucose concentrations of the respective culture mediums (see also Section 2.2) are given in protocols of the DSMZ (German Collection of Microorganisms and Cell Culture GmbH). Nevertheless, the highest PCR values of 6.50 ± 0.05 mV/min at 3.33 mM and 7.20 ± 0.06 mV/min at 5.0 mM were found with *L. brevis* bacteria. The plotted calibration curves in Figure 4 follow a kinetic behavior for a capacitive field-effect biosensor, which is exemplary discussed for *E. coli* in [46], similar like the Michaelis–Menten kinetics known for enzymatic reactions. Each calibration curve shows a saturation-like behavior with maximum PCR values (e.g., 7.20 ± 0.06 mV/min for *L. brevis* cells at 5.0 mM glucose), where no higher signal changes were detected when glucose concentration is increased further. In this context, two points might be taken into consideration: 1) after the acidification phase with higher cell numbers, the measured medium on the sensor surface will get more acidic. Due to a change of the physiological conditions in the acidic environment on the sensor surface, the activity of some cells in close proximity to the sensor surface might get blocked. Hence, a detection limit to higher PCR values can be caused; 2) in the measurement chambers (1–3), higher cell numbers in suspensions might inhibit the diffusion of more glucose molecules to the underlying cells on the pH-sensitive transducer layer.

In the next step, one glucose concentration (1.67 mM, blue arrow in Figure 4) was selected and all microorganisms were compared in terms of varying cell numbers (0.3×10^9, 0.6×10^9, 1.2×10^9, 2.4×10^9, and 4.8×10^9 cells). Here, five independent successive differential measurements were performed with the four-chamber LAPS set-up. Figure 5 shows the corresponding calibration curves with PCR values of *L. brevis*, *C. glutamicum*, and *E. coli* bacteria.

By increasing the cell number, the total metabolic response of the cells in the suspension increases. As explained for the calibration curves in Figure 4, a saturated-like behavior was observed, too. The highest PCR value of 5.60 ± 0.15 mV/min was obtained with *E. coli* cells at 4.8 × 10^9 cells. The lowest PCR value of 0.33 ± 0.04 mV/min was calculated for *L. brevis* at 0.3 × 10^9 cells. For the detailed values, see Table S3 in supplementary information of this article. The results from Section 3.2 show that for three different microorganisms, calibration matrices as function of glucose concentration or cell number can be defined. The obtained results are in good agreement with the results with single cell types on LAPS, which are reported in [27] for *E. coli* K12, in [29] for *C. glutamicum*, and in Section 3.1 for *L. brevis*.

3.3. Simultaneous Measurements with L. Brevis, C. Glutamicum, and E. Coli Bacteria

The extracellular acidification of *L. brevis*, *C. glutamicum*, and *E. coli* bacteria was determined simultaneously by means of the differential LAPS set-up. Here, the cell suspensions were parallelly prepared and then loaded into the chambers 1, 2, and 3. The fourth reference chamber was used again without cells. First, the four-chamber sensor arrangement was conditioned with 100 µL of the glucose concentration for 10 min. After this conditioning phase, 200 µL of respective cell numbers were added. In the first chamber *E. coli*, in the second chamber *C. glutamicum*, and in the third chamber *L. brevis* cell suspensions were pipetted. Three measurement repetitions were carried out to verify the reproducibility of all measurements.

For monitoring of the extracellular acidification, five independent, successive measurements (Nr. I–V) were performed (40 min each) by keeping the glucose concentration (1.67 mM) constant and varying the cell number (0.3 × 10^9, 0.6 × 10^9, 1.2 × 10^9, 2.4 × 10^9, 4.8 × 10^9 cells). Figure 6 depicts the LAPS signal responses before and after adding cells. After adding cells (10 min after starting each measurement, see diagrams, Nr. I–V), the sensor signal dropped due to an increase of the H^+-ion activity at the sensor surface.

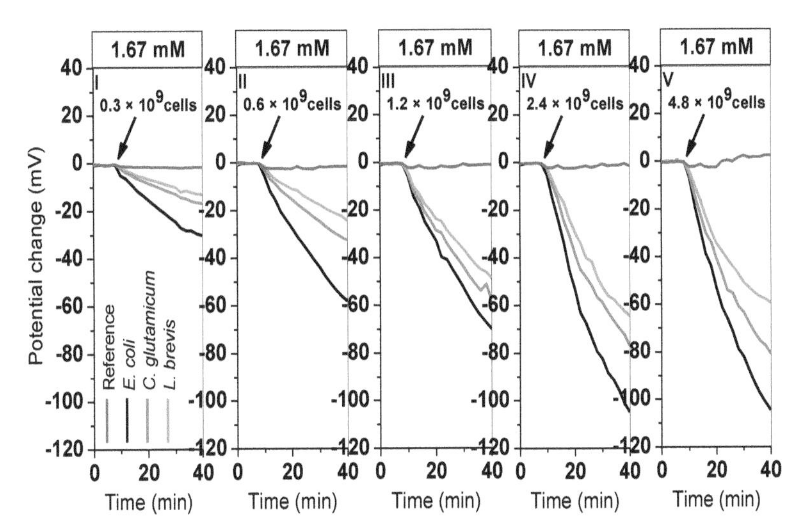

Figure 6. Four-chamber differential LAPS measurement with varying cell numbers of *L. brevis*, *C. glutamicum*, and *E. coli* (0.3 × 10^9, 0.6 × 10^9, 1.2 × 10^9, 2.4 × 10^9, 4.8 × 10^9 cells in 200 µL cell suspension) at a constant glucose concentration of 1.67 mM. The potential changes of five successive independent measurements (Nr. I–V) with an increasing number of cells are plotted. Blue line: reference sensor signal without cells; black (*E. coli*), red (*C. glutamicum*), and green (*L. brevis)* lines are active sensor areas with cells. Four laser-diode modules (LDMs) were considered for each chamber (each curve corresponds to the calculated mean values).

The signal changes are indicated with different colors (*E. coli* in black, *C. glutamicum* in red, *L. brevis* in green, and the reference signal in blue). In all set of measurements with different cell numbers, *E. coli* bacteria have shown the highest potential changes, followed by *C. glutamicum* and *L. brevis* cells. For instance, in the first set of measurements (I) and at the lowest cell number of 0.3×10^9 cells, a potential change of $\approx$ 30 mV (ΔpH $\approx$ 0.6) was observed with *E. coli* cells, following by C. *glutamicum* ($\approx$ 17 mV, ΔpH $\approx$ 0.32), and *L. brevis* ($\approx$ 13 mV, ΔpH $\approx$ 0.24); the values were calculated as described in Section 3.1. Correspondingly, in the last set of measurements (V) at the highest cell number of 4.8×10^9 cells, a potential change of $\approx$ 104 mV (ΔpH $\approx$ 1.93) was recorded with *E. coli*, $\approx$ 80 mV (ΔpH $\approx$ 1.5) with *C. glutamicum*, and $\approx$ 60 mV (ΔpH $\approx$ 1.11) for *L. brevis* cells. It was observed that by increasing the cell number, the extracellular acidification of bacteria was increased, as expected in agreement with the measurements where the acidification behavior of the cells was examined separately (see Sections 3.1 and 3.2). In the presence of higher cell numbers, more acids will be produced, which results in an increase of potential change values.

By the simultaneous evaluation with three different microorganisms on a single LAPS chip a kind of 'signal pattern' of the metabolic behavior of cells can be obtained. This way, it is possible to study them under absolute identical boundary conditions, which enables to have comparable results in real time. Furthermore, the measurement time can be reduced, which is advisable for time-critical cell-based measurements. Here, metabolic responses of three microorganisms were successfully evaluated within 40 min, whereas by sequential measuring in Section 3.1, 40 min was required for each model organism.

The resulting PCR values from Figure 6 were calculated and plotted as calibration curve in Figure 7, which follows a similar behavior, as it was explained for Figures 4 and 5.

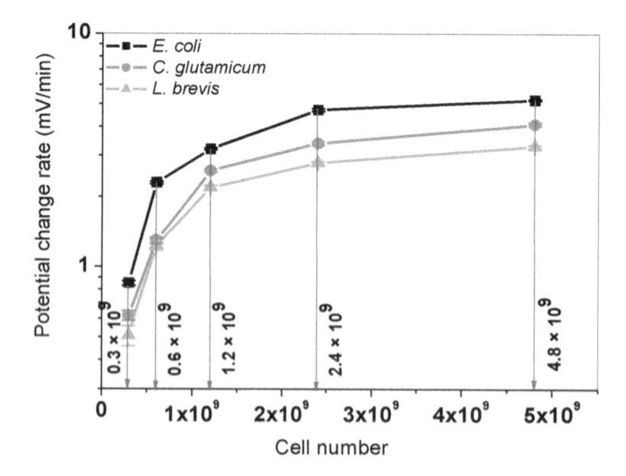

Figure 7. Correlation between the LAPS signal response and the cell number at a fixed glucose concentration of 1.67 mM. Mean values and standard deviations of PCR values of *E. coli* (black), *C. glutamicum* (red), and *L. brevis* (green) are depicted for different cell numbers (0.3×10^9, 0.6×10^9, 1.2×10^9, 2.4×10^9, 4.8×10^9 cells in 200 µL cell suspension). Three independent measurements were performed, and within the first 6 min after adding cells, the mean PCR values were calculated.

The highest PCR value of 5.20 ± 0.04 mV/min was calculated for *E. coli* bacteria for with 4.8×10^9 cells at 1.67 mM glucose. Respectively, the highest PCR values of 4.10 ± 0.06 mV/min (*C. glutamicum*), and 3.30 ± 0.05 mV/min (*L. brevis*) were obtained. The lowest detected PCR values were found at the lowest cell number (0.3×10^9 cells) and 1.67 mM glucose: 0.85 ± 0.02 mV/min (*E. coli*), 0.62 ± 0.03 mV/min (*C. glutamicum*), and 0.51 ± 0.05 mV/min (*L. brevis*). For the detailed values, see Table S4 in supplementary information of this article.

In order to validate whether different microorganisms in simultaneously performed measurements might mutually influence the signal characteristics of each other (cross-talk between four chambers), PCR values between the two experiments (sequential/simultaneous) for *E. coli*, *C. glutamicum*, and *L. brevis* are compared in Table 1; |Sequential PCR|, |Simultaneous PCR| and the difference between the two experiments |Δ PCR| are listed. Starting with a cell number of 0.3×10^9 cells up to 4.8×10^9 cells, signal value differences (|Δ PCR|) were studied: on the one hand, for most cell numbers, independent of the cell type, a good correlation was found with |Δ PCR| ≤ 0.46 mV/min, representing usual variations in preparation of cell suspensions and cultivation process. Such slight deviations in cell numbers will influence the metabolic response and consequently, the extracellular acidification, and finally the sensor output signal. On the other hand, the highest variation was obtained for *C. glutamicum* at 4.8×10^9 cells (0.8 mV/min) and at 1.2×10^9 cells (0.5 mV/min). A possible explanation for this discrepancy might be differences in surviving cell numbers after manual preparation of cell suspensions (in diluted PBS solution) after the cultivation process. Through automatization of the cultivation steps and parallelization of the sample preparation process, PCR value variations between measurements can be further minimized in future experiments. In addition, measurements have also shown that both experimental procedures (sequential/simultaneous measurement) are in good accordance with published data for *E. coli* and *C. glutamicum* bacteria [27,29].

Table 1. Comparison of mean values and standard deviations of the potential change rate values obtained from sequential measurements (Figure 5) and simultaneous measurements (Figure 7) after adding cells for *E. coli*, *C. glutamicum*, and *L. brevis*. The glucose concentration of 1.67 mM was kept constant and the cell number was varied from 0.3×10^9 up to 4.8×10^9 cells. |Δ PCR| describes the signal difference between both experiments.

Cells In 200 μL	Type Of Bacteria	|Sequential PCR| (mV/min)	|Simultaneous PCR| (mV/min)	|Δ PCR| (mV/min)
	E. Coli	0.84 ± 0.13	0.85 ± 0.02	0.01
0.3×10^9	*C. Glutamicum*	0.49 ± 0.05	0.62 ± 0.03	0.13
	L. Brevis	0.33 ± 0.04	0.51 ± 0.05	0.18
	E. Coli	1.90 ± 0.12	2.30 ± 0.03	0.40
0.6×10^9	*C. Glutamicum*	1.02 ± 0.04	1.30 ± 0.04	0.28
	L. Brevis	0.78 ± 0.06	1.24 ± 0.06	0.46
	E. Coli	3.10 ± 0.11	3.20 ± 0.03	0.10
1.2×10^9	*C. Glutamicum*	2.10 ± 0.04	2.60 ± 0.04	0.50
	L. Brevis	1.76 ± 0.05	2.20 ± 0.05	0.44
	E. Coli	4.50 ± 0.10	4.70 ± 0.06	0.20
2.4×10^9	*C. Glutamicum*	3.10 ± 0.22	3.40 ± 0.07	0.30
	L. Brevis	2.40 ± 0.06	2.80 ± 0.04	0.40
	E. Coli	5.60 ± 0.15	5.20 ± 0.04	0.40
4.8×10^9	*C. Glutamicum*	4.90 ± 0.04	4.10 ± 0.06	0.80
	L. Brevis	3.70 ± 0.02	3.30 ± 0.05	0.40

4. Conclusions

This article comprises two essential achievements in terms of LAPS-based biosensors: i) for the first time, a cell-based LAPS set-up was utilized to determine the extracellular acidification of *L. brevis* bacteria by varying the two parameters of glucose concentration and cell number. It was observed that by increasing the cell number and/or glucose concentration, the extracellular acidification of *L. brevis* cells increases as expected, due to an increase of H^+-ion activity on the sensor surface after the acidification phase. The overall sensor characteristic is comparable to data published for *E. coli* and *C. glutamicum*, however, the absolute values of its metabolic response differ, depending on the cell type. ii) For the first time, simultaneous measurements were carried out with three different microorganisms (*E. coli*, *C. glutamicum*, *L. brevis*) on the same LAPS chip. Hence, a 'signal pattern' of the extracellular acidification of cells was defined. There was no mutual influence on the signal characteristic for each type of bacteria. Moreover, a good correlation was found between the sequentially and simultaneously performed LAPS investigations.

In future studies, the differential LAPS set-up with model microorganisms can be applied to analyze the metabolic behavior of microorganism populations in (bio)fermentation broths (e.g., in a biogas fermentation broth). With the knowledge about their cellular metabolism obtained by LAPS, the sensor signal patterns with respective calibration curves can be defined. Subsequently, the metabolic 'interaction' between the fermentation broth and the particular model microorganisms can be evaluated: It can be examined, whether and how cells in the fermentation broth will influence the extracellular acidification of the analyzed model microorganisms. The related signal variations can be monitored at different process stages of a bioreactor, which might allow a better and faster control in case of process disturbances.

Supplementary Materials: The following are available online at http://www.mdpi.com/1424-8220/19/21/4692/s1, Table S1, Table S2, Table S3, Table S4.

Author Contributions: S.D. designed and performed the experiments, analyzed the data, and wrote the paper. D.R. helped by cell culturing. T.W., P.W., and M.J.S. supervised the work and reviewed the manuscript for the publication.

Funding: The authors thank the German Federal Ministry of Food and Agriculture (project no.: 22006613) and the German Federal Ministry of Education and Research (project no.: 13N12585) for the financial support of this work.

Acknowledgments: We would like to thank H. Iken for LAPS chip processing and B. Schneider for 3D-printed structures.

Conflicts of Interest: The authors declare no conflict of interest.

References

1. Macfarlane, G.T.; Macfarlane, S. Bacteria, colonic fermentation, and gastrointestinal health. *J. AOAC Int.* **2012**, *95*, 50–60. [CrossRef] [PubMed]
2. Devanthi, P.V.P.; Gkatzionis, K. Soy sauce fermentation: Microorganisms, aroma formation, and process modification. *Int. Food Res. J.* **2019**, *120*, 364–374. [CrossRef] [PubMed]
3. Haile, M.; Kang, W.H. The role of microbes in coffee fermentation and their impact on coffee quality. *J. Food Qual.* **2019**, *2019*, 1–6. [CrossRef]
4. Poszytek, K.; Pyzik, A. The effect of the source microorganisms on adaptation of hydrolytic consortia dedicated to anaerobic digestion of maize silage. *Anaerobe* **2017**, *46*, 46–55. [CrossRef] [PubMed]
5. O'Mara, P.; Farrell, A.; Bones, J.; Twomey, K. Staying alive! Sensors used for monitoring cell health in bioreactors. *Talanta* **2018**, *176*, 130–139. [CrossRef] [PubMed]
6. Reinecke, T.; Biechele, P.; Sobocinski, M.; Suhr, H.; Bakes, K.; Solle, D.; Jantunen, H.; Scheper, T.; Zimmermann, S. Continuous noninvasive monitoring of cell growth in disposable bioreactors. *Sens. Actuators B Chem.* **2017**, *251*, 1009–1017. [CrossRef]
7. Qui, J.; Arnold, M.A.; Murhammer, D.W. On-line near infrared bioreactor monitoring of cell density and concentrations of glucose and lactate during insect cell cultivation. *J. Biotechnol.* **2014**, *173*, 106–111.
8. Marquard, D.; Schneider-Barthold, C.; Düsterloh, S.; Scheper, T.; Lindner, P. Online monitoring of cell concentration in high cell density *Escherichia coli* cultivations using in situ microscopy. *J. Biotechnol.* **2017**, *259*, 83–85. [CrossRef]
9. Simmons, A.D.; Williams, C., III; Degoix, A.; Sikavitsas, V.L. Sensing metabolites for the monitoring of tissue engineered construct cellularity in perfusion bioreactors. *Biosens. Bioelectron.* **2017**, *90*, 443–449. [CrossRef]
10. Buchenauer, A.; Hofmann, M.C.; Funke, M.; Büchs, J.; Mokwa, W.; Schnakenberg, U. Micro-biosensors for fed-batch fermentations with integrated online monitoring and microfluidic devices. *Biosens. Bioelectron.* **2009**, *24*, 1411–1416. [CrossRef]
11. Hafeman, D.G.; Parce, J.W.; McConnell, H.M. Light-addressable potentiometric sensor for biochemical systems. *Science* **1988**, *240*, 1182–1185. [CrossRef] [PubMed]
12. Owicki, J.C.; Bousse, L.J.; Hafeman, D.G.; Kirk, G.L.; Olson, J.D.; Wada, H.G.; Parce, J.W. The light-addressable potentiometric sensor: Principles and biological applications. *Annu. Rev. Biophys. Biomol. Struc.* **1994**, *23*, 87–113. [CrossRef] [PubMed]

13. Wagner, T.; Schöning, M.J. Light-addressable potentiometric sensors (LAPS): Recent trends and applications. In *Electrochemical Sensor Analysis*; Alegret, S., Merkoci, A., Eds.; Elsevier: Amsterdam, The Netherlands, 2007; pp. 87–128.
14. Schöning, M.J. Playing around with field-effect sensors on the basis of EIS structures, LAPS and ISFETs. *Sensors* **2005**, *5*, 126–138. [CrossRef]
15. Yoshinobu, T.; Miyamoto, K.-I.; Werner, C.F.; Poghossian, A.; Wagner, T.; Schöning, M.J. Light-addressable potentiometric sensors for quantitative spatial imaging of chemical species. *Annu. Rev. Anal. Chem.* **2017**, *10*, 225–246. [CrossRef] [PubMed]
16. Das, A.; Yang, C.M.; Chen, T.C.; Lai, C.S. Analog micromirror-LAPS for chemical imaging and zoom-in application. *Vacuum* **2015**, *118*, 161–166. [CrossRef]
17. Miyamoto, K.-I.; Wagner, T.; Yoshinobu, T.; Kanoh, S.; Schöning, M.J. Phase-mode LAPS and its application to chemical imaging. *Sens. Actuator B Chem.* **2011**, *154*, 28–32. [CrossRef]
18. Chen, L.; Zhou, Y.; Jiang, S.; Kunze, J.; Schmuki, P.; Krause, S. High resolution LAPS and SPIM. *Electrochem. Commun.* **2010**, *12*, 758–760. [CrossRef]
19. Schöning, M.J.; Poghossian, A. Bio FEDs (field-effect devices), state-of-the-art and new directions. *Electroanalysis* **2006**, *18*, 1893–1900. [CrossRef]
20. Hizawa, T.; Sawada, K.; Takao, H.; Ishida, M. Fabrication of a two-dimensional pH image sensor using a charge transfer technique. *Sens. Actuators B Chem.* **2006**, *117*, 509–515. [CrossRef]
21. Dantism, S.; Takenaga, S.; Wagner, P.; Wagner, T.; Schöning, M.J. Determination of the extracellular acidification of *Escherichia coli* K12 with a multi-chamber-based LAPS system. *Phys. Status Solidi A* **2016**, *213*, 1479–1485. [CrossRef]
22. Sasaki, Y.; Kanai, Y.; Uchida, H.; Katsube, T. Highly sensitive taste sensor with a new differential LAPS method. *Sens. Actuators B Chem.* **1995**, *24–25*, 819–822. [CrossRef]
23. Wang, J.; Campos, I.; Wu, F.; Zhu, J.; Su, H.; Stellberg, G.B.; Palma, M.; Watkinson, M.; Krause, S. The effect of gold nanoparticles on the impedance of microcapsules visualized by scanning photo-induced impedance microscopy. *Electrochim. Acta* **2016**, *208*, 39–46. [CrossRef]
24. Yoshinobu, T.; Krause, S.; Miyamoto, K.-I.; Werner, C.F.; Poghossian, A.; Wagner, T.; Schöning, M.J. (Bio-) chemical Sensing and Imaging by LAPS and SPIM. In *Label-Free Biosensing*; Schöning, M.J., Poghossian, A., Eds.; Springer: Berlin/Heidelberg, Germany, 2018; pp. 103–132.
25. Werner, C.F.; Krumbe, C.; Schumacher, K.; Groebel, S.; Spelthahn, H.; Stellberg, M.; Wagner, T.; Yoshinobu, T.; Selmer, T.; Keusgen, M.; et al. Determination of the extracellular acidification of *Escherichia coli* by a light-addressable potentiometric sensor. *Phys. Status Solidi A* **2011**, *208*, 1340–1344. [CrossRef]
26. Werner, C.F.; Groebel, S.; Krumbe, C.; Wagner, T.; Selmer, T.; Yoshinubo, T.; Baumann, M.E.M.; Keusgen, M.; Schöning, M.J. Nutrient concentration-sensitive microorganism-based biosensor. *Phys. Status Solidi A* **2012**, *209*, 900–904. [CrossRef]
27. Dantism, S.; Röhlen, D.; Wagner, T.; Wagner, P.; Schöning, M.J. Optimization of cell-based multi-chamber LAPS measurements utilizing FPGA-controlled laser-diode modules. *Phys. Status Solidi A* **2018**, *215*, 1–8. [CrossRef]
28. Dantism, S.; Takenaga, S.; Wagner, T.; Wagner, P.; Schöning, M.J. Differential imaging of the metabolism of bacteria and eukaryotic cells based on light-addressable potentiometric sensors. *Elecrochim. Acta* **2017**, *246*, 234–241. [CrossRef]
29. Dantism, S.; Röhlen, D.; Selmer, T.; Wagner, T.; Wagner, P.; Schöning, M.J. Quantitative differential monitoring of the metabolic activity of *Corynebacterium glutamicum* cultures utilizing a light-addressable potentiometric sensor system. *Biosens. Bioelectron.* **2019**, *139*, 11332. [CrossRef]
30. Zhang, D.W.; Wu, F.; Wang, J.; Watkinson, M.; Krause, S. Image detection of yeast *Saccharomyces cerevisiae* by light-addressable potentiometric sensors (LAPS). *Elechtrochem. Commun.* **2016**, *72*, 41–45. [CrossRef]
31. Du, L.; Wang, J.; Chen, W.; Zhao, L.; Wu, C.; Wang, P. Dual functional extracellular recording using a light-addressable potentiometric sensor for bitter signal transduction. *Anal. Chim. Acta* **2018**, *1022*, 106–112. [CrossRef]
32. Liu, Q.; Hu, N.; Zhang, F.; Wang, H.; Ye, W.; Wang, P. Neurosecretory cell-based biosensor: Monitoring secretion of adrenal chromaffin cells by local extracellular acidification using light-addressable potentiometric sensor. *Biosens. Bioelectron.* **2012**, *35*, 421–424. [CrossRef]

33. Udaka, S. The discovery of *Corynebacterium glutamicum* and birth of amino acid fermentation industry in Japan. In *Corynebacteria: Genomics and Molecular Biology*; Burkovski, A., Ed.; Caister Academic Press: Norfolk, UK, 2008.
34. Rönkä, E.; Malinen, E.; Saarela, M.; Rinta-Koski, M.; Aarnikunnas, J.; Palva, A. Probiotic and milk technological properties of *Lactobacillus brevis*. *Int. J. Food Microbiol.* **2003**, *83*, 63–74. [CrossRef]
35. Schöning, M.J.; Brinkmann, D.; Rolka, D.; Demuth, C.; Poghossian, A. CIP (cleaning-in-place) suitable "non-glass" pH sensor based on Ta_2O_5-gate EIS structure. *Sens. Actuators B Chem.* **2005**, *111*, 423–429. [CrossRef]
36. Poghossian, A.; Schöning, M.J. Silicon-based chemical and biological field-effect sensors. In *Encyclopedia of Sensors*; Grimes, C.A., Dickey, E.C., Pishko, M.V., Eds.; American Scientific Publisher: Stevenson Ranch, CA, USA, 2006; pp. 463–534.
37. Schöning, M.J.; Arzdorf, M.; Mulchandani, P.; Chen, W.; Mulchandani, A. A capacitive field-effect sensor for the direct determination of organophosphorus pesticides. *Sens. Actuators B Chem.* **2003**, *91*, 92–97. [CrossRef]
38. Poghossian, A.; Thust, M.; Schroth, P.; Steffen, A.; Lüth, H.; Schöning, M.J. Penicillin detection by means of silicon-based field-effect structures. *Sens. Mater.* **2001**, *13*, 207–223.
39. Werner, C.F.; Wagner, T.; Yoshinubo, T.; Keusgen, M.; Schöning, M.J. Frequency behaviour of light-addressable potentiometric sensors. *Phys. Status Solidi A* **2011**, *210*, 884–891. [CrossRef]
40. Poghossian, A.; Werner, C.F.; Buniatyan, V.V.; Wagner, T.; Miyamoto, K.-I.; Yoshinubo, T.; Schöning, M.J. Towards addressability of light-addressable potentiometric sensors: Shunting effect of non-illuminated region and cross-talk. *Sens. Actuators B Chem.* **2017**, *244*, 1071–1079. [CrossRef]
41. Yoshinobu, T.; Ecken, H.; Poghossian, A.; Simonis, A.; Iwasaki, H.; Lüth, H.; Schöning, M.J. Constant-current-mode LAPS (CLAPS) for the detection of penicillin. *Electroanalysis* **2001**, *13*, 733–736. [CrossRef]
42. Poghossian, A.; Mai, D.-T.; Mourzina, Yu.; Schöning, M.J. Impedance effect of an ion-sensitive membrane: Characterization of an EMIS sensor by impedance spectroscopy, capacitance-voltage and constant-capacitance method. *Sens. Actuators B Chem.* **2004**, *103*, 423–428. [CrossRef]
43. Schöning, M.J.; Wagner, T.; Wang, C.; Otto, R.; Yoshinubo, T. Development of a handheld 16 channel pen-type LAPS for electrochemical sensing. *Sens. Actuators B Chem.* **2005**, *108*, 808–814. [CrossRef]
44. Wagner, T.; Rao, C.; Kloock, J.P.; Yoshinobu, T.; Otto, R.; Keusgen, M.; Schöning, M.J. "LAPS Card"—A novel chip card-based light-addressable potentiometric sensor (LAPS). *Sens. Actuators B Chem.* **2006**, *118*, 33–40. [CrossRef]
45. Schöning, M.J.; Näther, N.; Auger, V.; Poghossian, A.; Koudelka-Hep, M. Miniaturised flow-through cell with integrated capacitive EIS sensor fabricated at wafer level using Si and SU-8 technologies. *Sens. Actuators B Chem.* **2005**, *108*, 986–992. [CrossRef]
46. Huck, C.; Schiffels, J.; Herrera, C.N.; Schelden, M.; Selmer, T.; Poghossian, A.; Baumann, M.E.M.; Wagner, P.; Schöning, M.J. Metabolic responses of *Escherichia coli* upon glucose pulses captured by a capacitive field-effect sensor. *Phys. Status Solidi A* **2013**, *210*, 926–931. [CrossRef]

Article

Label-Free Detection of *E. coli* O157:H7 DNA Using Light-Addressable Potentiometric Sensors with Highly Oriented ZnO Nanorod Arrays

Yulan Tian [1,†], Tao Liang [2,†], Ping Zhu [1], Yating Chen [1], Wei Chen [1], Liping Du [1], Chunsheng Wu [1,*] and Ping Wang [2,*]

1 Institute of Medical Engineering, Department of Biophysics, School of Basic Medical Sciences, Xi'an Jiaotong University, Xi'an 710061, China; cnyulantian@stu.xjtu.edu.cn (Y.T.); jewel121@stu.xjtu.edu.cn (P.Z.); ytc20201011@xjtu.edu.cn (Y.C.); weixianyang@yeah.net (W.C.); duliping@xjtu.edu.cn (L.D.)

2 Biosensor National Special Laboratory, Key Laboratory for Biomedical Engineering of Ministry of Education, Department of Biomedical Engineering, Zhejiang University, Hangzhou 310027, China; cooltao@zju.edu.cn

* Correspondence: wuchunsheng@xjtu.edu.cn (C.W.); cnpwang@zju.edu.cn (P.W.)

† These authors contributed equally.

Received: 30 September 2019; Accepted: 9 December 2019; Published: 12 December 2019

Abstract: The detection of bacterial deoxyribonucleic acid (DNA) is of great significance in the quality control of food and water. In this study, a light-addressable potentiometric sensor (LAPS) deposited with highly oriented ZnO nanorod arrays (NRAs) was used for the label-free detection of single-stranded bacterial DNA (ssDNA). A functional, sensitive surface for the detection of *Escherichia coli* (*E. coli*) O157:H7 DNA was prepared by the covalent immobilization of the specific probe single-stranded DNA (ssDNA) on the LAPS surface. The functional surface was exposed to solutions containing the target *E. coli* ssDNA molecules, which allowed for the hybridization of the target ssDNA with the probe ssDNA. The surface charge changes induced by the hybridization of the probe ssDNA with the target *E. coli* ssDNA were monitored using LAPS measurements in a label-free manner. The results indicate that distinct signal changes can be registered and recorded to detect the target *E. coli* ssDNA. The lower detection limit of the target ssDNA corresponded to 1.0×10^2 colony forming units (CFUs)/mL of *E. coli* O157:H7 cells. All the results demonstrate that this DNA biosensor, based on the electrostatic detection of ssDNA, provides a novel approach for the sensitive and effective detection of bacterial DNA, which has promising prospects and potential applications in the quality control of food and water.

Keywords: DNA biosensor; ZnO nanorod arrays; LAPS; label-free detection; *E. coli*

1. Introduction

The detection of bacterial deoxyribonucleic acid (DNA) would have broad technological implications in areas ranging from biomedicine and the food industry, to environmental control [1–4]. For instance, the detection of *Escherichia coli* (*E. coli*) DNA is of great significance to the quality control of food and water. *E. coli* O157:H7 is one of the most dangerous foodborne pathogens, occurring in a variety of foods and water; it is able to cause hemorrhagic colitis, and leads to various symptoms, such as bloody diarrhea [5–8]. It is highly desirable to develop novel approaches for the detection of bacterial DNA in a label-free and cost-effective manner. For this reason, DNA biosensors have attracted increasing attention for their potential use in the label-free detection of bacterial DNA molecules. The cutting edge of the development of DNA biosensors is currently based on labelling strategies, that enable signal transductions and enhance the sensors' sensitivity [9–12]. However, these approaches, while promising, suffer from inherent limitations that severely restrict their applicability. Firstly, labelling strategies require some sort of label, which increases the time and cost of DNA detection. Secondly, the labelling process can complicate the DNA-device fabrication and the DNA detection process. Furthermore, the setup of labeling approaches often requires expensive and huge instruments that limit their practical applications, especially in-field applications.

Recent progress in pushing the limits mentioned above by using label-free strategies has proven to be an alternative emerging approach for DNA detection [13,14]. The fast development of micro- and nanofabricated devices has opened up an exciting realm for the development of a new generation of label-free DNA biosensors. For instance, the label-free detection of *E. coli* O157:H7 DNA has been reported using piezoelectric sensors and electrochemical biosensors [5,15–17]. More recently, field-effect devices (FEDs) have offered a promising direct electrical readout for label-free DNA detection [18–21]. DNA-FEDs detect the intrinsic molecular charges of DNA molecules. These devices are unrelated to any labelling for signal transduction. They have shown tremendous promise for label-free DNA detection in a much faster, more efficient manner. The basic mechanism is based on the electrical detection of surface-charge changes caused by the hybridization of probe single-stranded DNA (ssDNA) molecules (immobilized on the sensor's surface) with complementary target ssDNA molecules on the gate surface of the FEDs. As a result, the vast majority of DNA-FEDs are developed for the detection of ssDNA molecules. In this context, little is known about using FEDs for the direct label-free detection of bacterial ssDNA, which is a common and challenging task in many applications, such as the quality control of food and water [22,23].

The light-addressable potentiometric sensor (LAPS) is a type of FEDs. It has a light-addressable gate surface and is suitable for use as a transducer for the label-free electrical detection of DNA molecules based on their intrinsic molecular charges [21,24]. ZnO nanorods have shown promising potential in many applications, including sensors, due to their properties of having a direct wide bandgap, large exciton binding energy, and a high aspect ratio [25]. ZnO nanorods have the advantage of being highly oriented, well-structured, and being easy to prepare on substrate surfaces [26–28]. In addition, ZnO NRAs allow the loading of more functional sensing molecules due to their enlarged special surface area, which could potentially improve the sensing capability of the biosensors. Hence, it is essential to explore the feasibility of LAPS using ZnO NRA deposits for the direct label-free detection of bacterial DNA.

Inspired by this idea, a LAPS with ZnO NRA deposits, which is a typical FED with flexible gate electrodes, was used for the direct label-free electrostatic detection of negatively charged bacterial ssDNA molecules using their intrinsic molecular charges for the first time. The probe ssDNA molecules were covalently attached onto the LAPS surface via the silanization process. ZnO NRAs deposited on the LAPS surface could enlarge the surface area necessary for the probe ssDNA immobilization and provide three-dimensional (3D) sites for the probe ssDNA molecules to hybridize more effectively with the target ssDNA. The hybridization of the probe ssDNA with the target *E. coli* O157:H7 ssDNA was measured by recording the shifts in the LAPS photocurrent–voltage (*I–V*) curves. It is worth noting that the LAPS chips deposited with ZnO NRAs and bacterial ssDNA sequences used in this study only demonstrate the technical feasibility of this novel approach for direct bacterial ssDNA sensing.

2. Materials and Methods

2.1. Fabrication and Functionalization of LAPS Chip

A LAPS chip was fabricated using a p-doped silicon wafer with a thickness of 400 μm (<100>, 1–10 Ωcm). A 30 nm SiO_2 layer was developed on the surface of the silicon wafer through thermal oxidation. To create an Ohmic contact and light illumination, a 300 nm Al layer patterned with a window was fabricated on the rear side of the silicon wafer after etching the rear side of the SiO_2 layer. The hydrothermal method was employed to prepare a layer of ZnO NRAs on the surface of the LAPS chip as previous reported [26]. Briefly, a thin layer of ZnO film was deposited on the surface of the LAPS chip via reactive magnetron sputtering, which was used as the seeding layer. The hydrothermal deposition was performed in a solution with equal volumes of $Zn(NO_3)_2$ and methenamine at the same concentration of 0.02 M in a sealed beaker. The surface on which the ZnO NRAs were expected to grow was put downward at 95 °C for 2 h. Then, the LAPS chip was thoroughly rinsed with deionized water, and blown dry with N_2 for structure characterization and further sensing experiments.

The sequence of the probe ssDNA was designed specifically for the *E. coli* O157:H7 *eaeA* gene, which is a 30-base oligonucleotide (5′-AACGC CGATA CCATT ACTTA TACCG CGACG-3′). The sequences of the fully mismatched ssDNA were 5′-GCAGC GCCAT ATTCA TTACC ATAGC CGCAA-3′, which contains a fully mismatched base sequence to the probe ssDNA. Both the probe ssDNA and the fully mismatched ssDNA were synthesized by the Takara Biotechnology Company, Limited. To avoid any influence caused by solution changes on the measurement, 10 mM of phosphate-buffered saline (PBS) (pH 7.5) was used as the measurement solution for all the measurements and preparation of the probe ssDNA solutions. Next, 5′-end amino-modified probe ssDNA molecules were covalently immobilized on the LAPS surface via the silanization process. Briefly, the LAPS surface, deposited with ZnO NRAs, was treated with 0.1% (v/v) 3-aminopropyltriethoxysilane (APTES) in toluene with 1 h of incubation at room temperature (RT). Then, the LAPS surface was rinsed with the solvent and blown dry using N_2, then heated for 2 h at 120 °C to solidify the attachment. The introduced amine residues reacted with the added glutaraldehyde overnight at room temperature. The introduced aldehyde residues reacted with the amine residues on the end of the probe ssDNA at RT for 12 h. Finally, a 1% bovine serum albumin (BSA) was added to block any unreacted aldehyde residues and any other non-specific binding sites on the ZnO NRAs. The sensor was rinsed with PBS and stored at 4 °C for further experiments.

2.2. Preparation of Target *E. coli* O157:H7 ssDNA

The target ssDNA with part of the base sequence complementary to the probe ssDNA was prepared from the *E. coli* O157:H7 *eaeA* gene using an asymmetric polymerase chain reaction (PCR). First, the *E. coli* O157:H7 were cultured in a nutrient broth at 37°C for 12 h, then it was killed using a 100 °C water bath for 15 min. Then, the *E. coli* O157:H7 was serially diluted to the desired concentrations with PBS by the surface plating-count method. The *E. coli* O157:H7 genomic DNA was extracted using an E.N.Z.A Bacterial DNA Kit (Beijing Solarbio Science & Technology Co. Ltd, China) following the product's instructions and used as the DNA template for the asymmetric PCR in order to achieve the target *E. coli* O157:H7 ssDNA for label-free DNA detection, based on the hybridization of the target ssDNA with the probe ssDNA. The forward and reverse primers specific to the *E. coli* O157:H7 *eaeA* gene were 5′-GGCGG ATAAG ACTTC GGCTA-3′ and 5′-CGTTT TGGCA CTATT TGCCC-3′, respectively. For asymmetric PCR, the concentration of reverse primers was set to be 50 times higher than that of the forward primers. As a result, the forward primers played the role of a "limiting primer" (i.e., the target *E. coli* O157:H7 *eaeA* gene ssDNA will be generated by the reverse primer after the limiting forward primer was consumed). The asymmetric PCR reaction was carried out using a Bio-Rad Thermal Cycler (Bio-Rad Laboratories, Inc., Hercules, CA, USA) under the following conditions: 95 °C for 5 min preincubation, followed by 38 cycles of 30 sec denaturation at 95 °C, 30 sec of annealing at 55 °C, a 45 sec extension at 72 °C, and a 10 sec final extension. The product of the asymmetric PCR was a short ssDNA fragment (151 bases) of the *E. coli* O157:H7 *eaeA* gene, which contained the complementary base sequence to the probe ssDNA, thus allowing for the hybridization of the probe ssDNA and the target ssDNA on the LAPS surface for the detection of *E. coli* O157:H7 ssDNA.

2.3. LAPS Measurement Setup and Target ssDNA Detection

Figure 1 is a schematic diagram of the LAPS measurement setup used in this study, which is similar to those in our previous reports [29]. The light source was an He–Ne semiconductor laser (Coherent Co., Santa Clara, CA, USA) with a wavelength of 543.5 nm and a diameter of 1 mm for illuminating the local area on the LAPS chip. A potentiostat (EG & G Princeton Applied Research, M273A, USA) was used to provide bias voltage; this was applied to the LAPS chip for the generation of photocurrent during illumination. A lock-in amplifier (model SR830 DSP, Stanford Research Systems) was utilized to amplify the photocurrent of the LAPS chip. A National Instruments data-acquisition card (DAQmx PCI-6259, National Instruments, TX, USA) was employed to record and collect the amplified photocurrent. Home-made LabVIEW software was used to control the whole measurement setup. All measurements were carried out at RT. The whole setup was shielded with a Faraday box to exclude ambient light and to minimize the influences of environmental factors on the measurements.

The LAPS surface charges changed after the probe ssDNA hybridized with the target ssDNA on its surface, which can be detected by recording the shifts of the LAPS photocurrent–voltage curves (*I–V* curves). First, the measurement solution (10 mM PBS, pH 7.5) was added to the detection chamber to record the *I–V* curves from the LAPS chip functionalized with the probe ssDNA. Then, the measurement solution in the detection chamber was removed and incubated with a solution containing the target *E. coli* O157:H7 ssDNA for 1 h at RT. Then, the measurement solution was used to wash the detection chamber, to remove any ssDNA molecules that did not hybridize with the probe ssDNA on the LAPS surface. Finally, the *I–V* curves were recorded from the LAPS surface after the DNA hybridization. The charge changes induced by the target ssDNA hybridization with the probe ssDNA on the LAPS surface can be indicated by a comparison of the *I–V* curves recorded from the LAPS before and after the target ssDNA hybridization. The shifts of the *I–V* curves were calculated and used as an indicator of the target ssDNA hybridization that occurred on the LAPS surface functionalized with the probe ssDNA.

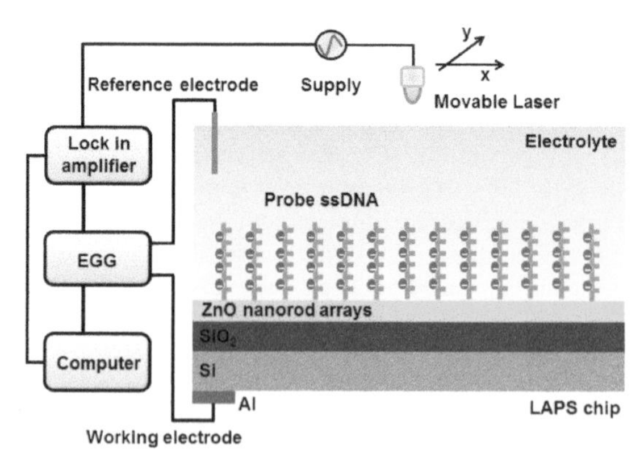

Figure 1. Schematic diagram of the light-addressable potentiometric sensor (LAPS) measurement setup.

3. Results and Discussion

3.1. Probe ssDNA Immobilization on the LAPS Surface

The surface of the LAPS chip used in this study was deposited with a layer of ZnO NRAs, which allowed us to load more probe ssDNA molecules due to their enlarged special surface area. A silanization process was performed to treat the LAPS surface and covalently immobilize the probe ssDNA on the sensor surface. A scanning electron microscope (SEM) was employed to characterize the ZnO NRAs deposited on the LAPS surface before and after silanization. Figure 2a shows the SEM image of the ZnO NRAs deposited on the LAPS surface before silanization. It can be observed that the orientation of the ZnO NRAs on the LAPS surface is highly ordered. The diameters of the ZnO NRAs are close to 50 nm, and their lengths and the space between the rods are about 500 nm and 100 nm, respectively. This geometry provided an ideal surface for the loading of more probe ssDNA. Figure 2b shows the SEM images of ZnO NRAs after silanization. It reveals that, with the treatment of 5% APTES, the attachment of APTES residues is apparent on the surface of the nanorods, thus providing sites for the further crosslink of the probe ssDNA using glutaraldehyde as bridges. Furthermore, after the silanization with APTES, the nanorods retained the same morphology, through which the probe ssDNA molecules could diffuse into and subsequently immobilize on the surface of the nanorods through a condensation reaction.

Figure 2. SEM images of ZnO nanorod arrays (NRAs) deposited on the LAPS surface (**a**) before and (**b**) after silanization with 3-aminopropyltriethoxysilane (APTES).

For the detection of the target *E. coli* O157:H7 ssDNA, it was necessary to functionalize the LAPS surface with the probe ssDNA, which contained a specific complementary base sequence to the target ssDNA. BSA blocking is also desirable to avoid the non-specific adsorption of molecules on the sensor surface. The LAPS surface modifications mentioned above can lead to shifts in the LAPS *I*–*V* curves. It is due to the surface charge changes originated from the attachment of charged molecules on the LAPS surface. Figure 3a clearly shows that both the probe ssDNA immobilization and the BSA blocking-induced shifts of the LAPS *I*–*V* curves to the positive direction of bias voltage. This is mainly due to the adsorption of the negatively charged probe ssDNA and BSA molecules on to the LAPS surface. In addition, the optimal concentration of probe ssDNA was found to be 5 μM, as indicated by the highest shifts of the LAPS *I*–*V* curves. This reflects that the probe ssDNA molecules are immobilized on the LAPS surface with greater efficiency. As compared to the LAPS chip without the ZnO NRAs, significant higher shifts of the LAPS *I*–*V* curves can be induced by probe ssDNA immobilization (Figure 2b). This proves our hypothesis that the LAPS chip with ZnO NRAs is able to load more probe ssDNA. This could improve the capability of the sensing target ssDNA, especially in increasing the sensor dynamic range. On the other hand, with regard to the shifts of LAPS *I*–*V* curves induced by BSA blocking, no obvious difference was found between the LAPS chips with and without ZnO NRAs (Figure 2b). All the results demonstrated that, compared to the LAPS chip without ZnO NRAs, the LAPS chip with ZnO NRAs was able to load more probe ssDNA molecules, and this could potentially enhance the detection of the target ssDNA.

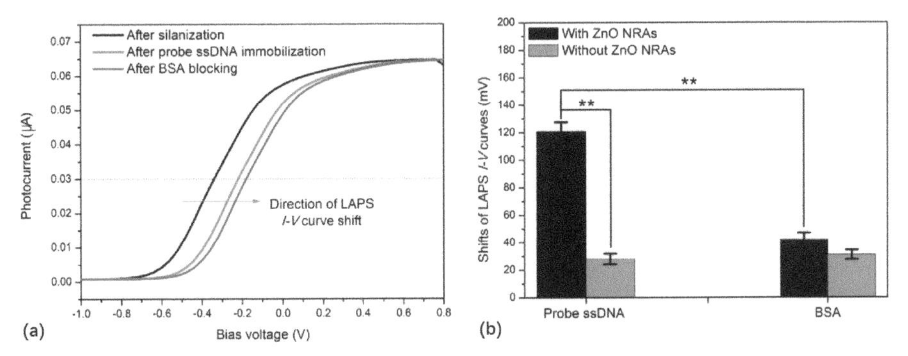

Figure 3. (**a**) *I*–*V* curves recorded from LAPS chip with ZnO NRAs after silanization, probe ssDNA immobilization, and bovine serum albumin (BSA) blocking. (**b**) Shifts of *I*–*V* curves induced by probe ssDNA immobilization and BSA blocking on the surface of LAPS chips with and without ZnO NRAs. All the data are represented by the mean ± standard error of the mean (SEM). ** $p < 0.01$, Student's *t*-test. The mean and SEM of three experiments are shown.

3.2. Dectection of Target E. coli O157:H7 ssDNA

After the probe ssDNA immobilization and BSA blocking, the prepared DNA biosensor was utilized for the detection of the target ssDNA amplified using an asymmetric PCR from different concentrations of *E. coli* O157:H7, ranging from 10 CFU/mL to 10^5 CFU/mL. Asymmetric PCR makes the PCR product become ssDNA, which makes it possible to hybridize it with the probe ssDNA, thus allowing it to be detected directly by the DNA biosensor. Moreover, the DNA concentrations of asymmetric PCR products are proportional to the concentrations of *E. coli* O157:H7. The proportional constant between the concentrations of target DNA molecules and *E. coli* O157:H7 was estimated to be 2.75×10^{11} according to the thermal cycles used in the protocol of asymmetric PCR. Therefore, the target ssDNA concentrations corresponding to 10 CFU/mL and 10^5 CFU/mL *E. coli* O157:H7 were estimated to be 4.57 nM and 45.7 μM, respectively. The measurement results indicate that asymmetric PCR products lead to the shifting of LAPS *I–V* curves to the more positive direction of bias voltage (Figure 4a). This is mainly attributed to the hybridization of the target ssDNA molecules with the probe ssDNA immobilized on the LAPS surface, which introduces more negative charges to the sensor surface, due to the molecules' intrinsic negative charges. In addition, more shifts of the LAPS *I–V* curves were observed to the positive direction of bias voltage when the asymmetric PCR products from higher concentrations of *E. coli* O157:H7 were applied for measurement. The shift of the LAPS *I–V* curves caused by the asymmetric PCR products from concentrations lower than 100 CFU/mL of *E. coli* O157:H7 were negligible, which indicates that the detection limit of this DNA biosensor for the detection of *E. coli* O157:H7 was as low as 100 CFU/mL.

The LAPS chip without ZnO NRAs was functionalized with the probe ssDNA and employed as a comparison to show the influences of ZnO NRAs on the performance of this DNA biosensor. As shown by the comparison in Figure 4b, it is obvious that the LAPS chip with ZnO NRAs showed a higher detection capability for *E. coli* O157:H7, as indicated by the higher slope of responsive linear curves. The detection limit of the LAPS chip without ZnO NRAs is 500 CFU/mL of *E. coli* O157:H7. This is probably because more probe ssDNA was attached to the LAPS chip with ZnO NRAs compared to the LAPS chip without ZnO NRAs. In addition, ZnO NRAs could provide 3D sites for the probe ssDNA molecules to hybridize with the target ssDNA more effectively. On the other hand, fully mismatched ssDNA alone was applied to test the specificity of this DNA biosensor. The results show that the fully mismatched ssDNA only induced negligible shifts in the LAPS *I–V* curves. This demonstrated a good specificity for the detection of target ssDNA amplified from the *E. coli* O157:H7.

However, since this DNA biosensor can only respond to target ssDNA, it is necessary to amplify the target ssDNA from *E. coli* O157:H7 using asymmetric PCR, which makes it impossible for this biosensor to detect *E. coli* O157:H7 directly from real samples. In further work, we will focus on the integration of sample preparation and treatment unit in this DNA biosensor, which could allow for the direct detection of real samples of *E. coli* O157:H7.

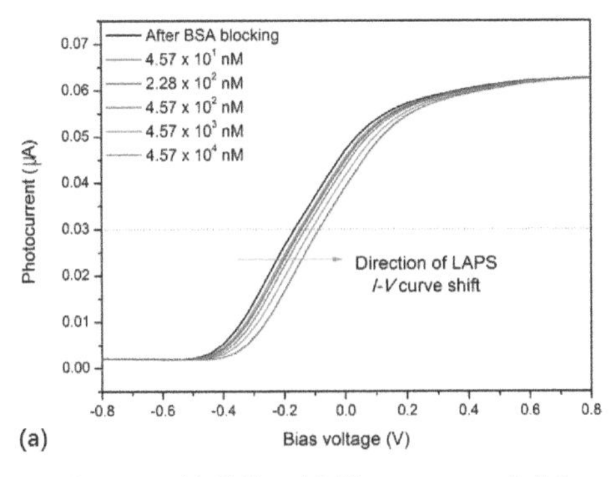

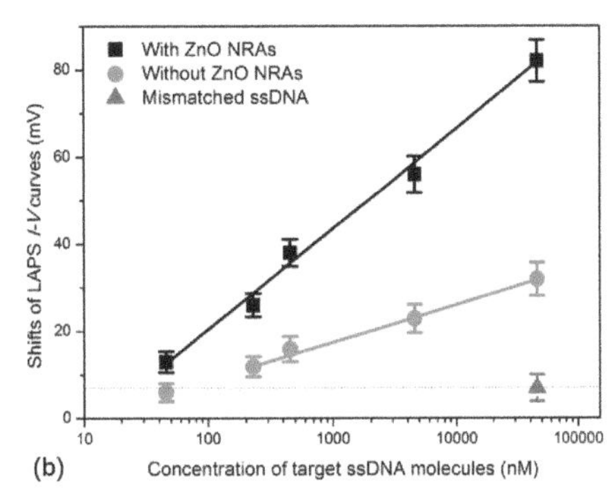

Figure 4. (**a**) Shifts of *I–V* curves recorded from the LAPS chip with ZnO NRAs in response to asymmetric polymerase chain reaction (PCR) products from different concentrations of *E. coli* O157:H7. (**b**) Statistical results of potential shifts of *I–V* curves recorded from LAPS chips with and without ZnO NRAs, induced by target ssDNA amplified from different concentrations of *E. coli* O157:H7. All the data are represented by the mean ± standard error of the mean (SEM). The mean and SEM of six experiments are shown.

4. Conclusions

In this study, we have demonstrated that, with ZnO NRAs deposited on the surface of a LAPS, the hybridization of the target ssDNA amplified from *E. coli* O157:H7—with the probe ssDNA immobilized on the ZnO NRAs—can be detected by the shift of the *I–V* curve on the LAPS readout. Compared to the LAPS chip without ZnO NRAs, the sensing capability of DNA biosensors using a LAPS chip with ZnO NRAs improved significantly, as indicated by the higher responsive signals and lower detection limit. This improvement is mainly attributed to the dimensional and sensing properties of ZnO NRAs. In the near future, we will focus on the direct detection of ssDNA without amplification by integrating functional units to this DNA biosensor, which has great potential for the development of portable instruments toward the in-field detection of bacteria. With its ability to detect the hybridization of DNA molecules and its light-addressable characteristics, LAPS is a potential candidate for a new kind of label-free addressable DNA microarray and DNA chip, which has the advantages of saving time and decreasing the cost of potential applications in many fields, such as biomedicine, food and water quality control, and individualized medicine.

Author Contributions: Conceptualization, C.W. and P.W.; methodology, T.L. and Y.T.; software, Y.T.; validation, Y.C., P.Z. and T.L.; formal analysis, P.Z.; investigation, T.L.; resources, P.W.; data curation, W.C. and L.D.; writing—original draft preparation, Y.T.; writing—review and editing, C.W.; supervision, C.W. and P.W.; project administration, C.W. and P.W.; funding acquisition, L.D., C.W. and P.W.

Funding: This research was funded by the National Natural Science Foundation of China, China (Grant Nos. 51861145307, 31700859, and 31661143030), the Doctoral Fund of Education Ministry of China, China (Grant No. 2018M633524), and the Fundamental Research Funds for the Central Universities, China.

Conflicts of Interest: The authors declare no conflict of interest. The funders had no role in the design of the study; in the collection, analyses, or interpretation of data; in the writing of the manuscript, or in the decision to publish the results.

References

1. Teles, F.R.R.; Fonseca, L.P. Trends in DNA biosensors. *Talanta* **2008**, *77*, 606–623. [CrossRef]
2. Sassolas, A.; Leca-Bouvier, B.D.; Blum, L.J. DNA biosensors and microarrays. *Chem. Rev.* **2008**, *108*, 109–139. [CrossRef] [PubMed]
3. Zhao, W.W.; Xu, J.J.; Chen, H.Y. Photoelectrochemical DNA Biosensors. *Chem. Rev.* **2014**, *114*, 7421–7441. [CrossRef] [PubMed]

4. Zulkifli, S.N.; Rahim, H.A.; Lau, W.J. Detection of contaminants in water supply: A review on state-of-the-art monitoring technologies and their applications. *Sens. Actuators B Chem.* **2018**, *255*, 2657–2689. [CrossRef]
5. Wang, L.J.; Liu, Q.J.; Hu, Z.Y.; Zhang, Y.F.; Wu, C.S.; Yang, M.; Wang, P. A novel electrochemical biosensor based on dynamic polymerase-extending hybridization for *E. coli* O157:H7 DNA detection. *Talanta* **2009**, *78*, 647–652. [CrossRef]
6. Ariffin, E.Y.; Lee, Y.H.; Futra, D.; Tan, L.L.; Abd Karim, N.H.; Ibrahim, N.N.N.; Ahmad, A. An ultrasensitive hollow-silica-based biosensor for pathogenic Escherichia coli DNA detection. *Anal. Bioanal. Chem.* **2018**, *410*, 2363–2375. [CrossRef]
7. Ye, W.W.; Chen, T.; Mao, Y.J.; Tian, F.; Sun, P.L.; Yang, M. The effect of pore size in an ultrasensitive DNA sandwich-hybridization assay for the Escherichia coli O157:H7 gene based on the use of a nanoporous alumina membrane. *Microchim. Acta* **2017**, *184*, 4835–4844. [CrossRef]
8. Chan, K.Y.; Ye, W.W.; Zhang, Y.; Xiao, L.D.; Leung, P.H.M.; Li, Y.; Yang, M. Ultrasensitive detection of *E. coli* O157:H7 with biofunctional magnetic bead concentration via nanoporous membrane based electrochemical immunosensor. *Biosens. Bioelectron.* **2013**, *41*, 532–537. [CrossRef]
9. Chen, Y.; Wang, Q.; Xu, J.; Xiang, Y.; Yuan, R.; Cha, Y.Q. A new hybrid signal amplification strategy for ultrasensitive electrochemical detection of DNA based on enzyme-assisted target recycling and DNA supersandwich assemblies. *Chem. Commun.* **2013**, *49*, 2052–2054. [CrossRef]
10. Hu, R.; Liu, T.; Zhang, X.B.; Huan, S.Y.; Wu, C.C.; Fu, T.; Tan, W.H. Multicolor Fluorescent Biosensor for Multiplexed Detection of DNA. *Anal. Chem.* **2014**, *86*, 5009–5016. [CrossRef]
11. Golub, E.; Niazov, A.; Freeman, R.; Zatsepin, M.; Willner, I. Photoelectrochemical Biosensors Without External Irradiation: Probing Enzyme Activities and DNA Sensing Using Hemin/G-Quadruplex-Stimulated Chemiluminescence Resonance Energy Transfer (CRET) Generation of Photocurrents. *J. Phys. Chem. C* **2012**, *116*, 13827–13834. [CrossRef]
12. Zhang, C.Y.; Yeh, H.C.; Kuroki, M.T.; Wang, T.H. Single-quantum-dot-based DNA nanosensor. *Nat. Mater.* **2005**, *4*, 826–831. [CrossRef] [PubMed]
13. Ngo, H.T.; Wang, H.N.; Fales, A.M.; Vo-Dinh, T. Label-Free DNA Biosensor Based on SERS Molecular Sentinel on Nanowave Chip. *Anal. Chem.* **2013**, *85*, 6378–6383. [CrossRef] [PubMed]
14. Mertens, J.; Rogero, C.; Calleja, M.; Ramos, D.; Martín-Gago, J.A.; Briones, C.; Tamayo, J. Label-free detection of DNA hybridization based on hydration-induced tension in nucleic acid films. *Nat. Nanotechnol.* **2008**, *3*, 301–307. [CrossRef] [PubMed]
15. Wang, L.J.; Wei, Q.S.; Wu, C.S.; Hu, Z.Y.; Ji, J.; Wang, P. The *E. coli* O157:H7 DNA detection on a gold nanoparticle enhanced piezoelectric biosensor. *Chin. Sci. Bull.* **2008**, *53*, 1175–1184. [CrossRef]
16. Minaei, M.E.; Saadati, M.; Najafi, M.; Honari, H. Label-free, PCR-free DNA Hybridization Detection of Escherichia coli O157:H7 Based on Electrochemical Nanobiosensor. *Electroanalysis* **2016**, *28*, 2582–2589. [CrossRef]
17. Bahşi, Z.B.; Büyükaksoy, A.; Ölmezcan, S.M.; Şimşek, F.; Aslan, M.H.; Oral, A.Y. A Novel Label-Free Optical Biosensor Using Synthetic Oligonucleotides from *E. coli* O157:H7: Elementary Sensitivity Tests. *Sensors* **2009**, *9*, 4890–4900. [CrossRef]
18. Abouzar, M.H.; Poghossian, A.; Pedraza, A.M.; Gandhi, D.; Ingebrandt, S.; Moritz, W.; Schoening, M.J. An array of field-effect nanoplate SOI capacitors for (bio-) chemical sensing. *Biosens. Bioelectron.* **2011**, *26*, 3023–3028. [CrossRef]
19. Sorgenfrei, S.; Chiu, C.Y.; Gonzalez, R.L.; Yu, Y.J.; Kim, P.; Nuckolls, C.; Shepard, K.L. Label-free single-molecule detection of DNA-hybridization kinetics with a carbon nanotube field-effect transistor. *Nat. Nanotechnol.* **2011**, *6*, 126–132. [CrossRef]
20. Stine, R.; Robinson, J.T.; Sheehan, P.E.; Tamanaha, C.R. Real-Time DNA Detection Using Reduced Graphene Oxide Field Effect Transistors. *Adv. Mater.* **2010**, *22*, 5297–5300. [CrossRef]
21. Wu, C.S.; Bronder, T.; Poghossian, A.; Werner, C.F.; Schoening, M.J. Label-free detection of DNA using a light-addressable potentiometric sensor modified with a positively charged polyelectrolyte layer. *Nanoscale* **2015**, *7*, 6143–6150. [CrossRef] [PubMed]
22. Branquinho, R.; Veigas, B.; Pinto, J.V.; Martins, R.; Fortunato, E.; Baptista, P.V. Real-time monitoring of PCR amplification of proto-oncogene c-MYC using a Ta2O5 electrolyte–insulator–semiconductor sensor. *Biosens. Bioelectron.* **2011**, *28*, 44–49. [CrossRef] [PubMed]

23. Zhu, L.; Zhao, R.J.; Wang, K.G.; Xiang, H.B.; Shang, Z.M.; Sun, W. Electrochemical Behaviors of Methylene Blue on DNA Modified Electrode and Its Application to the Detection of PCR Product from NOS Sequence. *Sensors* **2008**, *8*, 5649–5660. [CrossRef] [PubMed]
24. Wu, C.S.; Poghossian, A.; Bronder, T.; Schoening, M.J. Sensing of double-stranded DNA molecules by their intrinsic molecular charge using the light-addressable potentiometric sensor. *Sens. Actuators B Chem.* **2016**, *229*, 506–512. [CrossRef]
25. Parthangal, P.M.; Cavicchi, R.E.; Zachariah, M.R. A universal approach to electrically connecting nanowire arrays using nanoparticles—Application to a novel gas sensor architecture. *Nanotechnology* **2006**, *17*, 3786–3790. [CrossRef]
26. Guo, M.; Diao, P.; Cai, S. Hydrothermal growth of well-aligned ZnO nanorod arrays: Dependence of morphology and alignment ordering upon preparing conditions. *J. Solid State Chem.* **2005**, *178*, 1864–1873. [CrossRef]
27. Zhang, Z.X.; Yuan, H.J.; Zhou, J.J.; Liu, D.F.; Luo, S.D.; Miao, Y.M.; Gao, Y.; Wang, J.X.; Liu, L.F.; Song, L.; et al. Growth Mechanism, Photoluminescence, and Field-Emission Properties of ZnO Nanoneedle Arrays. *J. Phys. Chem. B* **2006**, *110*, 8566–8569. [CrossRef]
28. Ahsanulhaq, Q.; Umar, A.; Hahn, Y.B. Growth of aligned ZnO nanorods and nanopencils on ZnO/Si in aqueous solution: Growth mechanism and structural and optical properties. *Nanotechnology* **2007**, *18*, 115603. [CrossRef]
29. Zong, X.L.; Wu, C.S.; Wu, X.L.; Lu, Y.F.; Wang, P. A non-labeled DNA biosensor based on LAPS modified with TiO_2 thin film. *J. Zhejiang Univ. Sci. B* **2009**, *10*, 860–866. [CrossRef]

MDPI
St. Alban-Anlage 66
4052 Basel
Switzerland
Tel. +41 61 683 77 34
Fax +41 61 302 89 18
www.mdpi.com

Sensors Editorial Office
E-mail: sensors@mdpi.com
www.mdpi.com/journal/sensors

Lightning Source UK Ltd.
Milton Keynes UK
UKHW051906091020
371310UK00007B/135

9 783039 430284